Entertaining
CLUB DESIGN
A Private Feast

私享盛宴：娱乐会所设计

精品文化工作室 / 编　　康琪　王连双　张海云 / 译

大连理工大学出版社

Da ian University of Technology Press

图书在版编目(CIP)数据

私享盛宴:娱乐会所设计:汉英对照/精品文化
工作室编;康琪,王连双,张海云译. — 大连:大连理
工大学出版社,2011.11
　　ISBN 978-7-5611-6563-8

　　Ⅰ.①私… Ⅱ.①精…②康…③王…④张… Ⅲ.
①休闲娱乐—服务建筑—建筑设计—图集 Ⅳ.
①TU247-64

　　中国版本图书馆CIP数据核字（2011）第203829号

出版发行：大连理工大学出版社
　　　　　（地址：大连市软件园路80号　邮编：116023）
印　　　刷：精一印刷（深圳）有限公司
幅面尺寸：246mm×290mm
印　　　张：23
插　　　页：4
出版时间：2011年11月第1版
印刷时间：2011年11月第1次印刷
责任编辑：刘　蓉
责任校对：李　雪
封面设计：四季设计

ISBN 978-7-5611-6563-8
定　　　价：328.00元

电　话：0411-84708842
传　真：0411-84701466
邮　购：0411-84703636
E-mail：designbooks_dutp@yahoo.cn
URL：http://www.dutp.cn

如有质量问题请联系出版中心：（0411）84709246　84709043

Entertaining
CLUB DESIGN
A Private Feast

Contents 目录

CLUB 会所

NIGHT CLUB　夜店

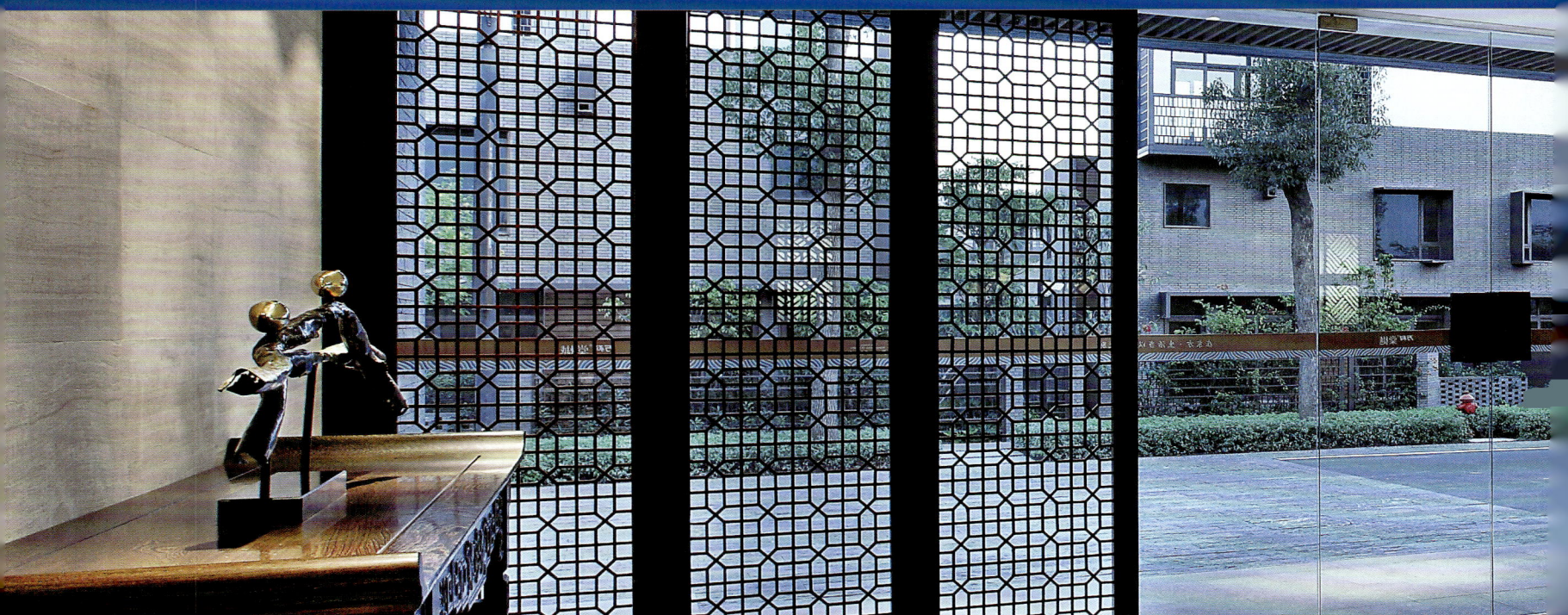

CLUB 会所

TONGREN SHANG ZUO CLUBHOUSE, GUIZHOU

贵州铜仁上座会馆

—— 地点
贵州

—— 面积
7000平方米

—— 设计师
熊华阳

—— 设计公司
深圳华空间机构

—— 主要材料
大理石、绿可木、布艺沙发、装饰画

A PRIVATE FEAST
Entertaining Club
DESIGN

私享盛宴：娱乐会所设计

Shangzuo Clubhouse is bonded together by two three-storied Chinese style architecture, which provides service sectors like body-building club, entertainment bar, Chinese-style catering and so on. With the geographical advantage, the designer makes full use of the local landscape of mountains and water and the characteristics of ethnic minorities as well. From exterior design to interior design, he incorporates Chinese design framework into modern-style furniture and accessories. The small pond in the middle of the courtyard, frescos with ethnic minorities characteristics, modern-style sofa, Chinese-style classical wooden chairs, carpets with bamboo leaves pattern, all these exhibit the traditional elegance and rich artistic conception of the clubhouse from the interior to the exterior. The interior design complements the products design and the two echo one another from one point to the whole space. With meticulous work, the designer creates a comfortable and pleasing clubhouse. In here, not only life is enjoyed, but art as well.

上垄会馆由两座三层高的中式建筑砌合而成，内含有健身会所、娱乐酒吧、中式餐饮等项目。设计师结合其地理优势，充分利用当地的山水之景及少数民族特色，从外观设计到室内设计，选用中式的设计框架结合现代风格的家具、饰品。院子中央的小池塘、少数民族特色的壁画、现代风格的沙发、中式古典的木椅、竹叶图案的地毯……使会所由内及外地散发出传统的高雅及丰富的意境。室内设计和产品设计相辅相成，由点及面的互相呼应，在设计师的精心打造下，创造出一个舒适惬意的会馆。在这里，不只是享受生活，更是享受艺术。

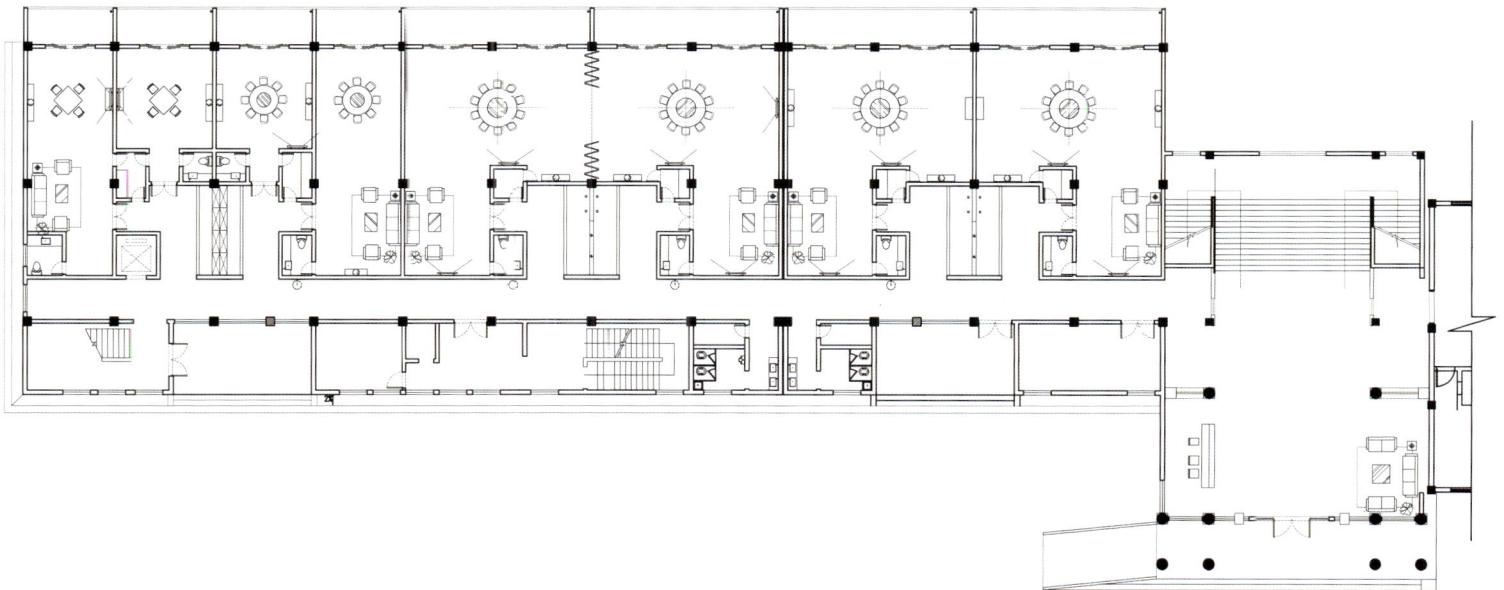

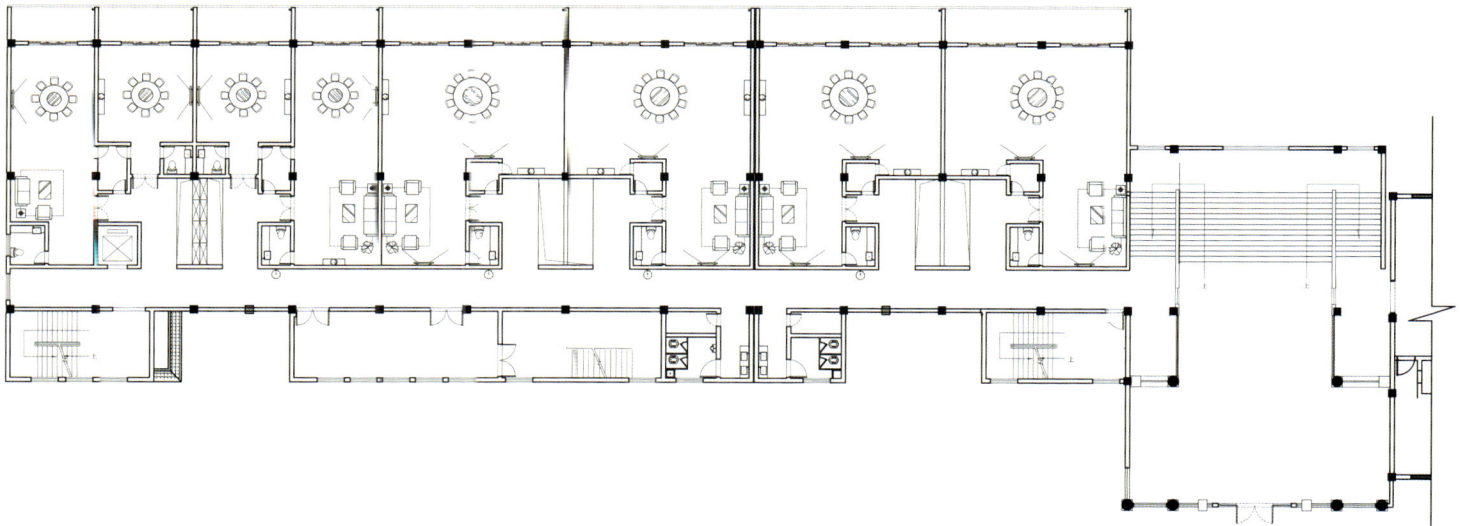

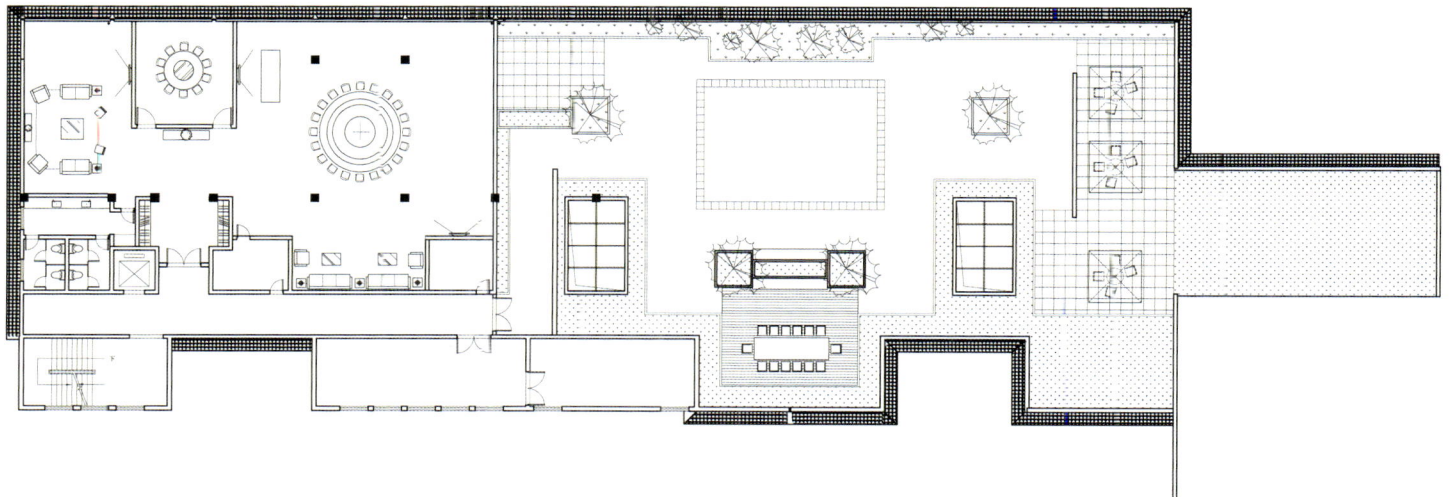

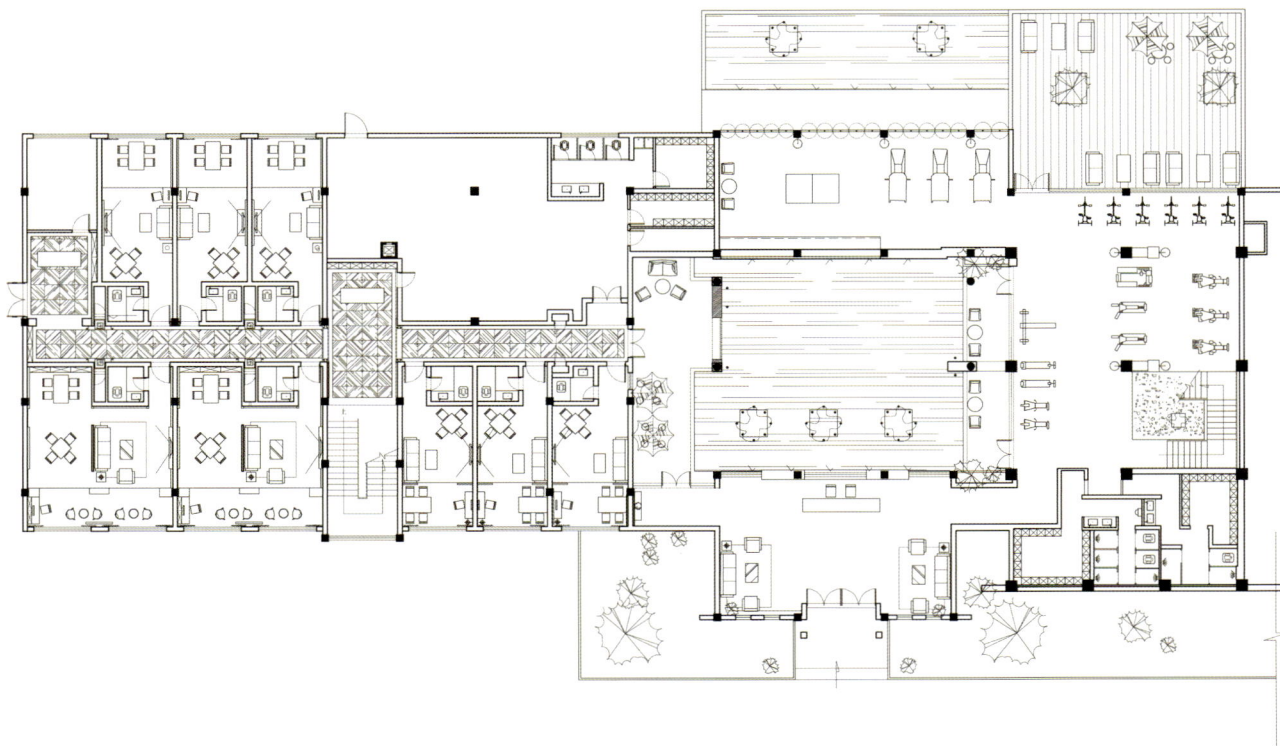

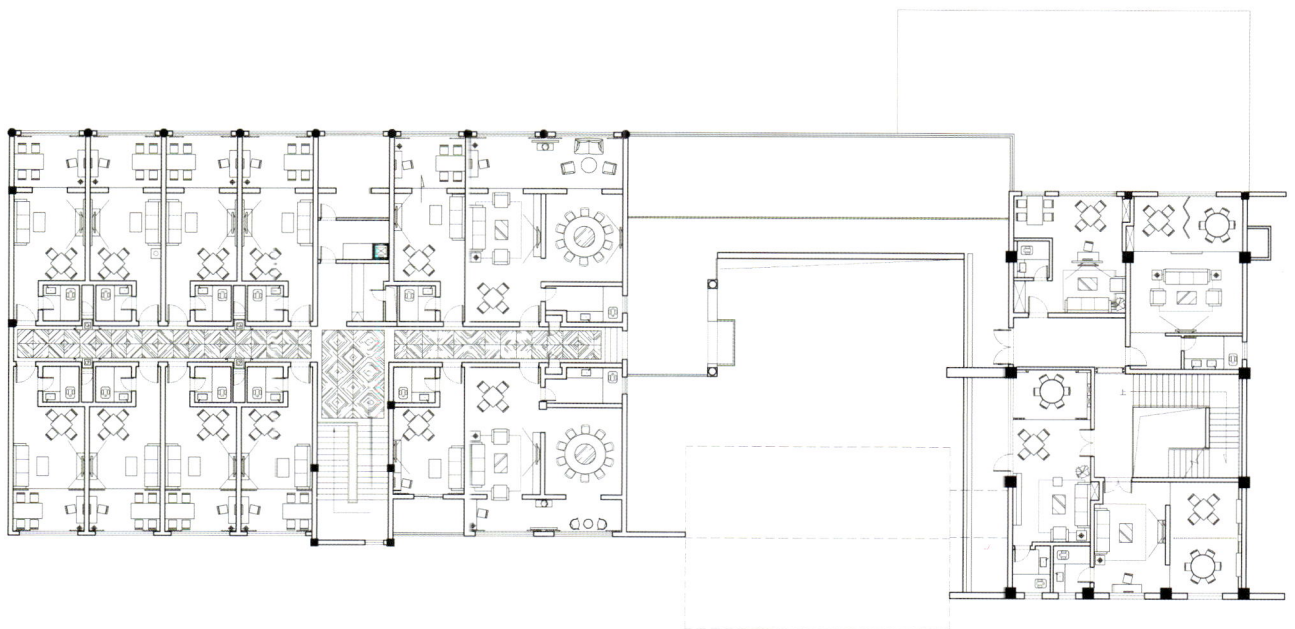

CHANGJIA CLUBHOUSE, SHANGHAI

长甲上海会所

— 地点

上海

— 设计公司

上海乐尚装饰设计工程有限公司

乐尚设计成立于1999年，是一家从事高档楼盘会所、售楼处、样板房、别墅、软装配饰与施工于一体的专业室内设计公司。乐尚人以"创意自由、规划严谨"的理念，将创意和规划优势进行整合，最大限度地为客户创造价值。

A PRIVATE FEAST
Entertaining Club
DESIGN

私享盛宴：娱乐会所设计

In the design of the clubhouse, the designers build a classical British-style space with sophisticated skills. No matter the outside appearance of the house or the interior decoration, both of them exemplify the traditional British style, which is elegant and with a taste of originality. From the classical nostalgic bar counters to the separated elegant booths, from the quiet corridors to the graceful space upstairs, the high ceiling, the complicated beam column structure... it's a place where people's memory of time and space fails when they become intoxicated in the magnificent feelings and become reluctant to go home. This is the charm of design and you can neither monotonously put the furniture pieces together, nor overlap the gracefulness. But it will be excellent if it suits the place and the article, matches the surroundings and serves for delight. And from now on, the comfortable enjoyment has just begun.

在这个会所设计中，设计师运用纯熟的手法打造了一个极具英式古典风情的空间。不论是外观建筑，还是室内装修，都体现出浓厚的英伦风情，优雅而别具意味。从古典、怀旧气息浓厚的吧台到独立、优雅的卡座，从幽静的走廊到典雅的楼上空间，有挑高的穹顶，有繁复的梁柱结构……让人忘记时间，忘记地点，只想沉醉在美妙的感觉中，不想归去。这就是设计的魅力，不是单调的家具拼凑，也不是优雅的重叠，只是适地适物，宜景怡情，舒适的享受，从此刻才刚刚开始。

SHENZHEN VANKE TANGYUE CLUBHOUSE

深圳万科棠樾会所

— 地点
东莞

— 面积
3000平方米

— 设计师
韩松

— 设计公司
深圳市昊泽空间设计有限公司

— 主要材料
木纹灰云石、黑木纹板、柚木、日本纸

A PRIVATE FEAST
Entertaining Club
DESIGN

私享盛宴：娱乐会所设计

In the design of this project, the designer employs traditional Chinese-style elements and classic skills to create a Chinese-style clubhouse which is full of cultural atmosphere. Walking in the clubhouse, you can feel the elegant and agreeable artistic teahouse, recall the state of mind when the wind blows through the bamboo grove, and experience the intimacy and ease that the Chinese-style home space brings. Because it is a public space, the selection of the furniture tends to be elegant and formal which highlights Chinese-style elements and carries Chinese-style culture. Log, of course, can't be omitted, while furniture reproductions of Ming and Qing Dynasties and small accessories like blue and white porcelain should be exquisite but not too much, which adds the vital finishing touch to the space. The lighting of the space is mostly soft and bright which can reduce the seriousness of the Chinese-style furnishings and allow the space to be more elegant and comfortable.

本案设计中，设计师运用中式传统元素与经典手法为人们营造出一个文化氛围浓重的中式风格会所。行走在会所里，感受清雅、惬意的茶艺馆，回味风过竹林的意境，体验中式家居空间带给人的亲切、自然。因为是公共空间，所以在家具的选择上侧重于雅致、大气的，能突出中式元素并饱含中式意蕴，原木材质自然不能少，仿清明家具陈设及青花瓷等小型装饰，重在精而不在多，在空间里起着画龙点睛的作用。空间的照明多温和、明亮，减轻中式风格带给人的沉重感，让空间更显清雅、惬意。

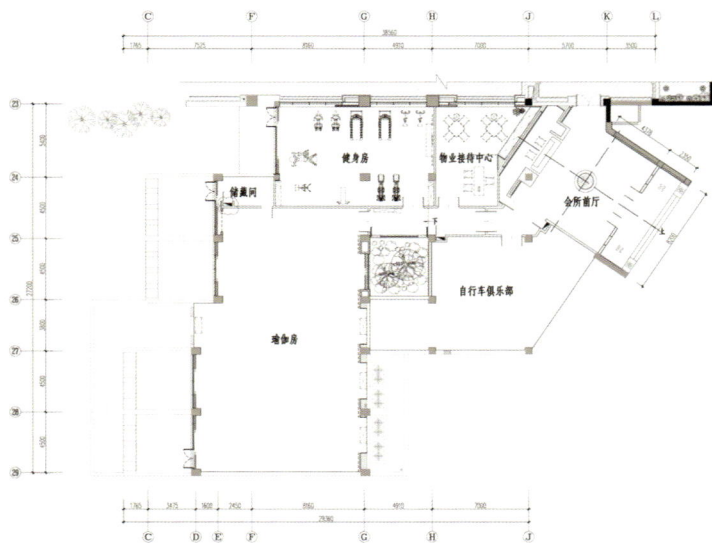

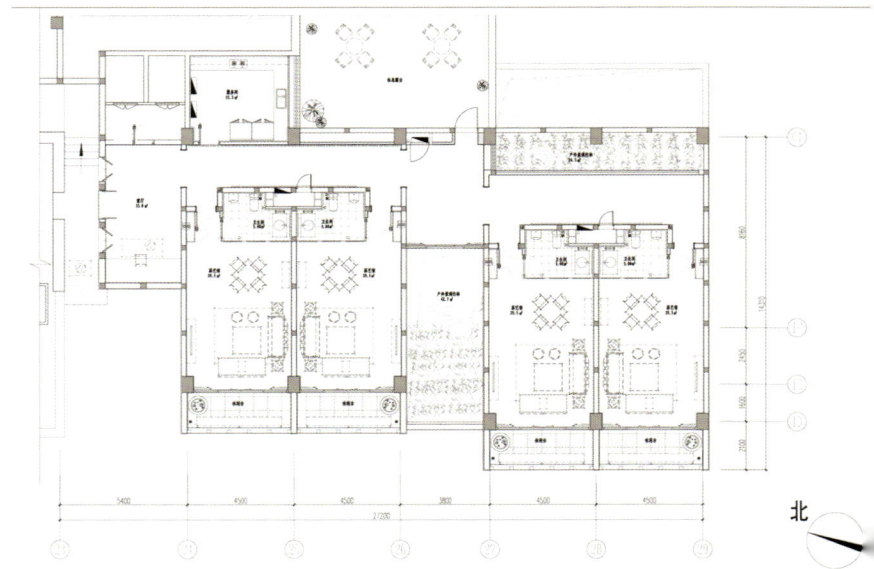

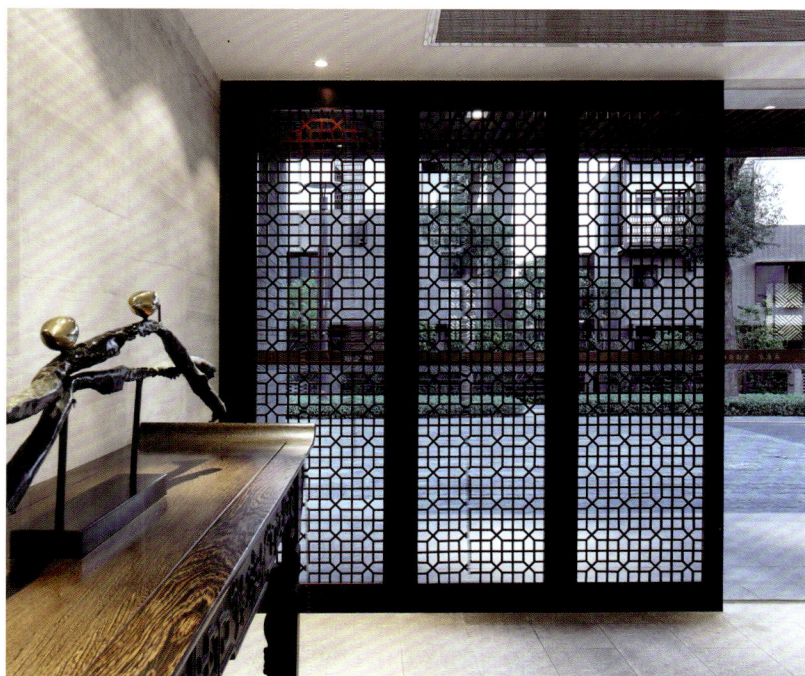

IMPERIAL COURT

御膳皇庭

— 地点
宜昌

— 面积
2400平方米

— 设计师
王治、范辉

— 设计公司
武汉艾亿威装饰设计顾问有限公司

— 主要材料
石材、地毯、墙纸、金箔、樱桃木、镜钢、皮料、玻璃

A PRIVATE FEAST
Entertaining Club
DESIGN

私享盛宴：娱乐会所设计

The design of this project is based on how to guide the high-end consumption and create a grand atmosphere of luxury and noble feeling. As for the design style, it adopts the approach of combining the east and west elements and mix-and-match, which both expresses the traditional culture and meets the modern aesthetic needs. By means of space division abundant in Chinese philosophical meaning and the decorative language throughout west and east, it conveys rich cultural feelings, while maintaining a relaxed atmosphere.

The clubhouse consists of the elevator hall, main hall, compartments, passageway and multi-functional hall. The space is mainly a private one, and all the rooms should be booked beforehand. Nine compartments with multiple functions form the main body of the clubhouse. The compartments cover an area of one hundred to two hundred square meters. All six compartments are interconnected to be the multi-functional hall which can also be used individually. Among the compartments are several passageways and hallways, emphasizing the sense of luxury and privacy.

本案的设计基础是如何引导高端消费，创造隆重奢华的氛围和尊贵的感受。设计风格上采用了中西结合和混搭的手法，既要表达传统文化又要符合现代审美需求。通过充满中式哲学意味的空间划分与融贯中西的装饰语言处理来传递丰富的文化感受，同时不失轻松的氛围。

会所由电梯厅、大厅、包间、通道和多功能厅组成。大多数空间都是私密空间，所有的房间都需要预定，九间功能齐全的包间构成了会所的主体。包间的面积在100平方米到200平方米之间。六间连通包间是可以化整为零的多功能厅。包间之间穿插了若干的通道和过厅，强调空间的豪华和私密感。

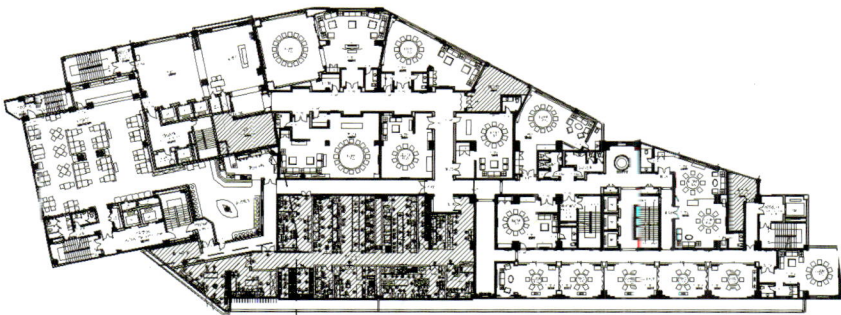

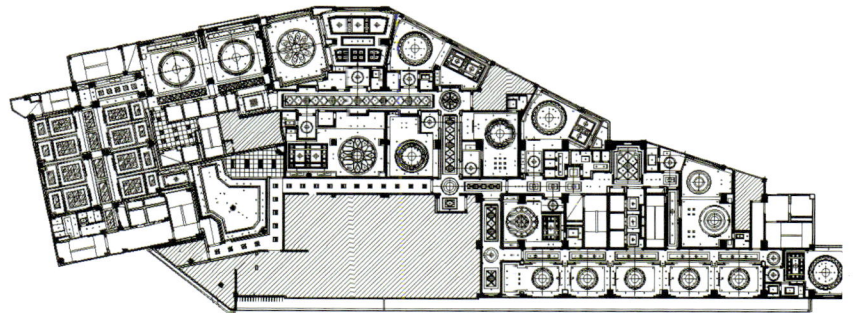

BINGO BILLIARDS PROVINCIAL SPORTS CENTER BRANCH

宾格台球省体店

— 地点
福州

— 面积
800平方米

— 设计师
高雄、高宪铭

— 设计公司
道和设计机构

— 主要材料
黑钛、水曲柳、黑镜、木质花格、木纹砖、蒙古
黑火烧拉槽板、墙纸、PVC管、地毯、绿可木

— 摄影师
周跃东

A PRIVATE FEAST
Entertaining Club
DESIGN

私享盛宴; 娱乐会所设计

The design of the front desk at the entrance and the cue lockers on both sides boldly use geometric cutting methods. The stoving varnished PVC pipes provide a sharp visual contrast for the background. With the embedded logo of the project, it sketches out modern geometric shapes. The wall surface of the recreation and waiting area is obliquely spliced with laminate board, which extends the space and serves as a partition. Not only does it meet the needs of the space, but also enhances the interactions of light and shadow inside and outside the compartments. The private tutoring area is a place which is enclosed by two sets of leisurely-used bar counters. Although it's called "private tutoring", it is not unapproachable. The wall surface of the left aisle follows the uneven facet modeling (wooden) of the front desk with showcases dotted within it.

On the design of line shape and block surface, manliness and robustness serve as the leading line throughout the whole space. The expression of light and shadow orients the style to be a modern, fashionable and recreational billiards club. The whole set of recreational place presents a beauty of sharp contrast, and you cannot refrain from having a try even if you are not a billiard player.

入口的前台以及两侧的寄杆柜大胆地将几何切割法运用其中、PVC管烤漆处理让背景从视觉上有了强烈的对比，再将本案的标志嵌入其中，勾勒出现代几何形体。休闲等候区的墙面由金刚板斜拼组成，使空间得到延伸，并且起到了隔断的作用，不仅满足了空间需求，也增强了包厢内外的光影互动。私教区是由两组可供休闲使用的吧台围合而成的，虽称之为"私教"，却并不拒人千里，左边过道的墙面沿用了前台的凹凸切割面造型（木作），将展示柜分散在其中。

设计线形和块面上以阳刚、硬朗为引线，贯穿了整个空间。用光影的诉说，将格调定位为现代时尚的娱乐台球会所。整套娱乐场所呈现出一种强烈的对比反差美，即使你不是球友，也难免跃跃欲试。

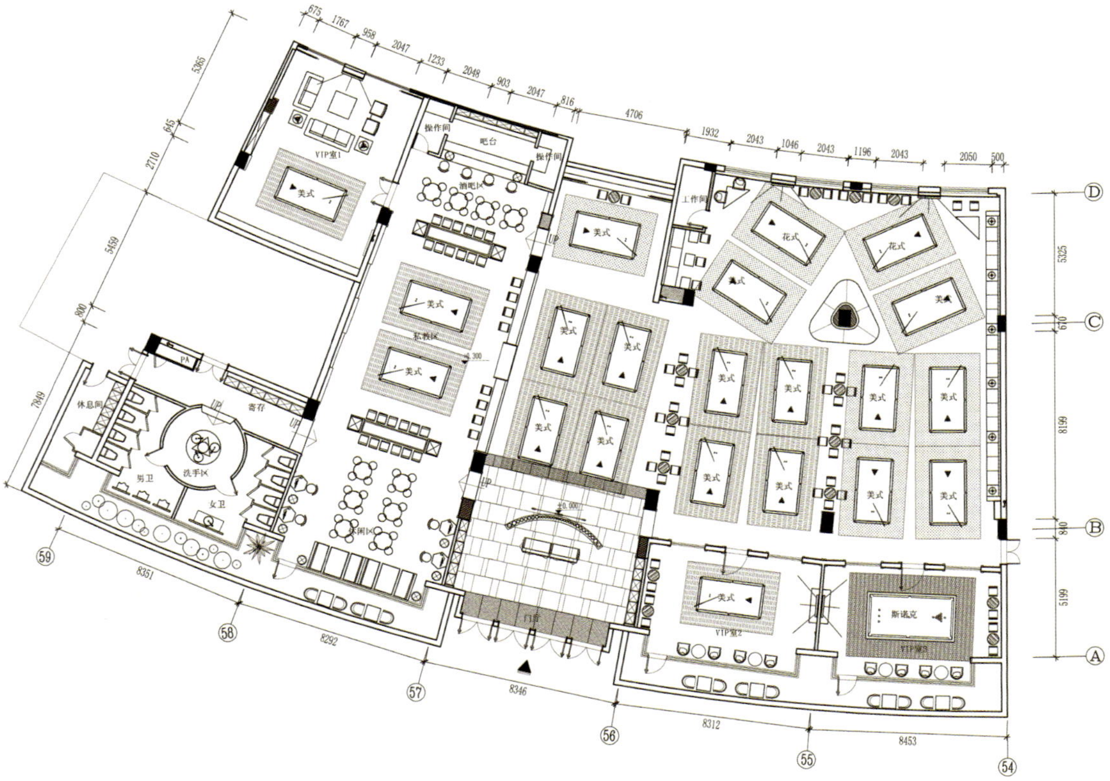

675
1767
958
2047
1233
2048
903
2047
816
4706
1932
2043
1046
2043
1196
2043
2050
500
5365
645
2710
5459
800
7849
5325
670
8199
840
5199
VIP室1
美式
操作间
吧台
操作间
酒吧区
私教区
美式
美式
UP
休息间
PA
寄存
洗手区
男卫
女卫
休闲区
±0.000
门厅
工作间
花式
花式
美式
美式
VIP室2
斯诺克
59
8351
58
8292
57
8346
56
8312
55
8453
54
D
C
B
A

MUFUXIANG BATH CLUBHOUSE

沐福祥沐浴会所

— 地点
  浙江
— 面积
  5000平方米
— 设计师
  王敏杰
— 设计公司
  翡冷翠设计咨询有限公司
— 主要材料
  大理石、玻璃、瓷砖、马赛克

A PRIVATE FEAST
Entertaining Club
DESIGN

私享盛宴; 娱乐会所设计

The designer bases his design on the principle of symmetry, which makes the elegant atmosphere echo the green landscape outdoors, and in this way, it creates a scene of extraordinary charm and beauty. When guests first enter the atrium, a nine-meter-high glass wall comes into sight, with an outdoor landscape pool as its background. Through the skylight, sunshine rests on the surfaces of different materials and textures, blending the harmonious atmosphere both indoors and outdoors, and creating a far-reaching perspective in the space. The atrium is a shared space, which adopts the centralized style, unified while varied. Other functional spaces stretch out in all directions, combining motion with quietness, and illusion with reality. The relationship between providing service and being served is developed to the extreme. The hue is mainly set in scarlet and off-white, with scarlet indicating little festivity and luxury, while off-white offering the customers visual relaxation and relieving customers of total fatigue, which makes them feel more sense of belonging. As for the lighting arrangement, with the transition from the public space to the private one, lighting changes from the brightness of natural sunshine to the dimness of artificial illumin of the space.

设计师的设计基础为对称原则，让高雅氛围与室外绿化景观呼应，营造出别有洞天的意境。首先进入中庭，印入眼帘的是9米高的落地玻璃，背景为户外景观池，阳光透过天窗洒落在不同材质、肌理上，交织出室内外的和谐氛围，让空间产生一种深远的透视感。中庭属于共享空间，采用中央集中式风格，统一中有变化，而其他功能空间则向四周展开，动静、虚实结合，让服务与被服务的关系达到极致。色彩基调为脂红色和米白色，红色带点喜庆和豪华，米白则使人视觉放松，让客户洗去一身的疲惫，更有归属感。在照明处理上，由公共空间转向私密空间时，灯光也由自然光的亮向人工照明的暗转变，加强了空间上的视觉层次。

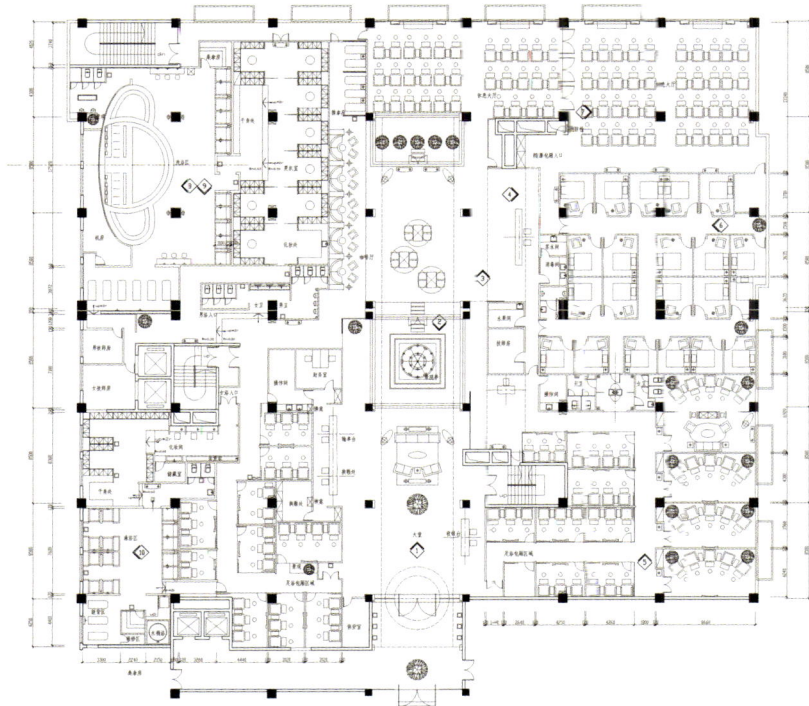

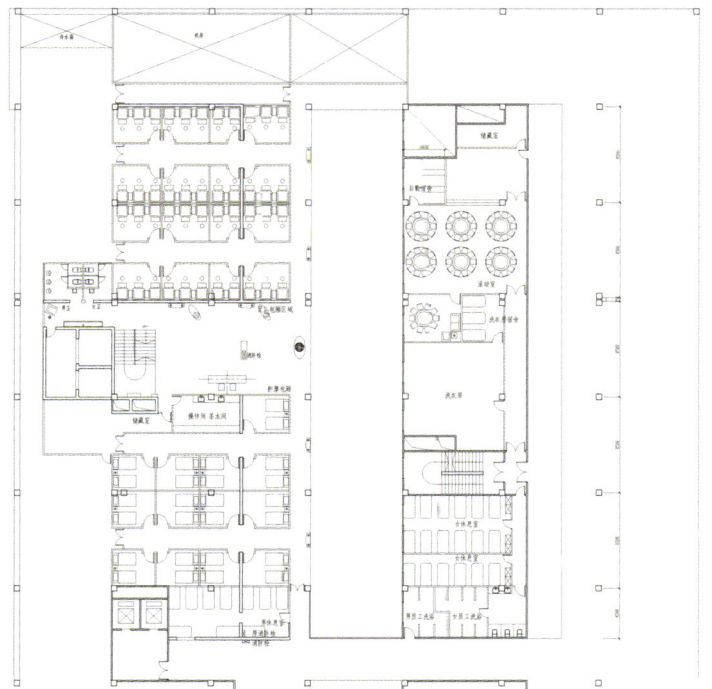

SHANYUANFANG TEA CLUBHOUSE

善缘坊茶会所

— 地点
福州

— 面积
500平方米

— 设计师
张开旺、林文

— 设计公司
张开旺设计事务所

— 主要材料
水曲柳面板深色、金钱花大理石、青石板火烧
面、条形青砖

A PRIVATE FEAST
Entertaining Club
DESIGN

私享盛宴：娱乐会所设计

This is a typical Chinese-structured tea bar, and the simple and elegant appearance leaves us the first impression of the tea clubhouse. The green bamboo at the front door adds bits of quietness to the environment, and the lighting makes the whole building exquisite and elegant.

The project is divided into two floors, the first one of which is mainly set for receiving guests, while the other serves mainly as the VIP compartments. At the entrance there are both the front reception area and reception service area, which provide the waiting service. Walking along the corridor, you will see ordinary compartments and dining rooms on one side and office rooms and stairwell on the other side. At the very end, a VIP room comes into sight. Beyond it there is a courtyard with the most traditional patio, demonstrating the original residential flavor. Meanwhile, the spacious and bright space guarantees the indoor natural lighting. There are three VIP rooms on the second floor, with the accordingly set bathrooms and storerooms, to offer quiet and convenient service to the guests. Marble, grey granite with blast-burn treated surface throughout the space and the traditional Chinese-style partition door are used to create a quiet and tranquil atmosphere. Together with the bar-shaped grey bricks in the rooms, the elegant Zenish ambiance is revealed to make people feel calm and peaceful.

这是一个典型中式结构的茶吧，简单清雅的外观是茶会所给人的第一印象，门前的绿竹为环境增添了几分清幽，灯光的辉映让整个建筑显得玲珑雅致。

本案共有两层，一层多用于接待，二层主要是VIP包厢。在入口处有接待前区和接待服务区，提供等候服务。沿着走道向前，两边分别是普通包厢、餐厅、办公区域和楼梯间，最后面是一个VIP房间，房间后面是一个内院，拥有最传统的天井，体现出原始的民居风味，同时敞亮的空间也保证了室内的采光。二楼有三个VIP房间，还有配套的卫生间和仓库，为客人创造宁静而方便的服务。贯穿始终的大理石、青石板火烧面及中式传统槅门，营造了宁静清幽的氛围，加上房间里的条形青砖，更是流露出清雅的禅意，让人安定、平和。

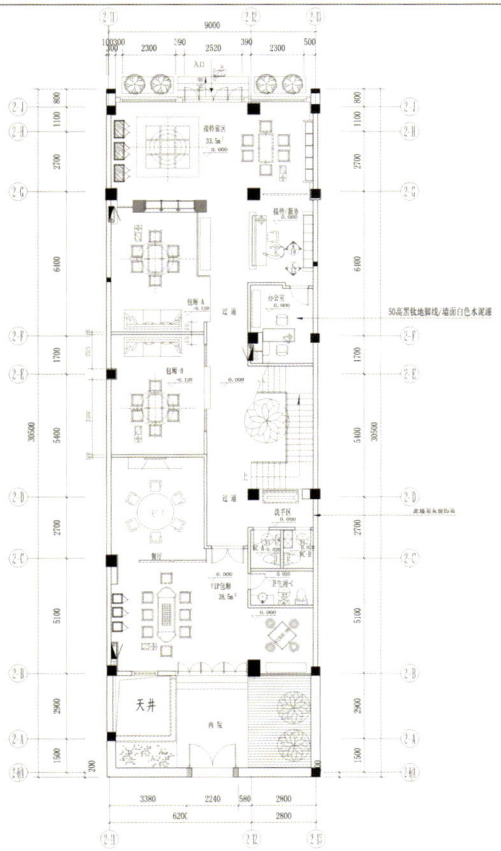

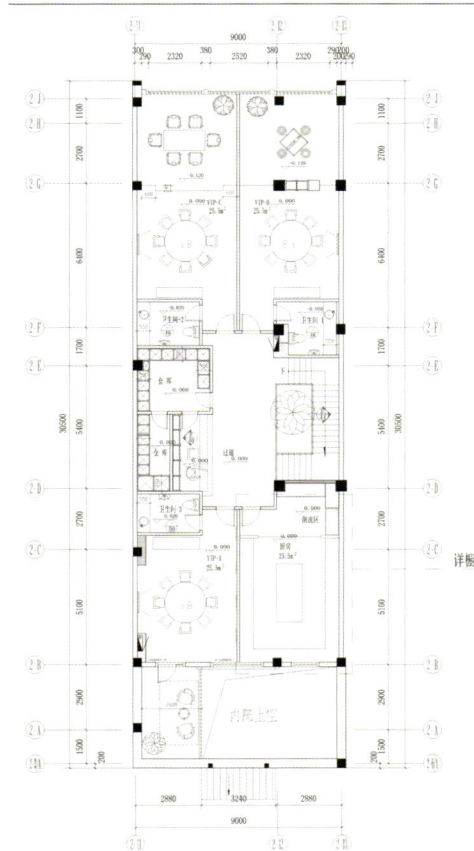

LIHAI SOVEREIGN BUILDING RESIDENT CLUBHOUSE

利海君临天下住户会所

— 地点

广州

— 面积

约1000平方米

— 设计师

史鸿伟、彭征

— 设计公司

广州共生形态工程设计有限公司

— 主要材料

木地板、黑色铁艺、不锈钢、墙纸、软包、地毯

A PRIVATE FEAST
Entertaining Club
DESIGN

私享盛宴：娱乐会所设计

As a detached clubhouse of the whole project, its interior design divides the building into several different functional areas including entertainment room, gym room, children's reading room, and piano gallery, etc.. Each functional area carries a distinctive style, being independent but also in harmony with the other ones to meet people's different psychological needs. In addition, aiming at target customers, the designers create various space experiences within the same overall environment to meet their demands for seeking fresh and stimulating sensory experience. The bright color matching, the bold and exaggerated shapes and the different lighting treatment…, all of them make the space special and novel, and enable the aesthetically tired and bored people in city life to feel revitalized enthusiasm and enjoy the wonderful life.

作为楼盘的独栋住户会所，室内设计将建筑分割成若干个不同的功能空间，包括娱乐室、健身房、儿童阅览室、钢琴廊等等。每个功能空间都有着各自不同的风格，各自独立又相互协调，满足人们不同的心理需求。此外，设计师还针对目标客户群，创造出多种不同的空间体验，在同一个大环境中，满足人们寻求新鲜与刺激的感官体验。绚丽的色彩搭配、大胆夸张的造型、不同的灯光处理……这一切都让空间充满个性和新奇，让在都市生活中产生审美疲劳和厌倦情绪的人们再次燃起热情，享受美好的生活！

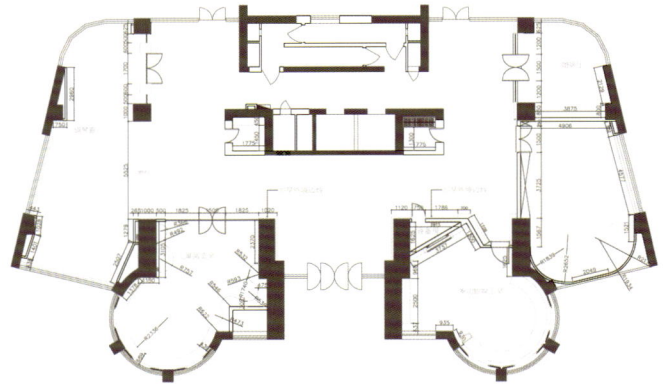

CLEAR WATER LUXURY GARDEN INTERNATIONAL WATER CLUB

碧水豪园国际水会

— 地点
宁波

— 面积
30000平方米

— 设计师
温浙平

— 设计公司
锐格建筑与室内设计事务所

— 主要材料
大理石、马赛克、绿可木、木饰面、镜面不锈钢、钢化玻璃

A PRIVATE FEAST
Entertaining Club
DESIGN

私享盛宴：娱乐会所设计

Costing 120 million yuan and covering a business area of 30,000 square meters, the Clear Water Luxury Garden International Water Club is a super-scaled public recreation project. In addition to the super-scaled male and female water spa area, 600-square-meter outdoor swimming pool and the waterfront viewing terrace, under the suggestion of RIG, for the first time the investors in the business introduce the Leisure Mall mode, and the well-known leisure brands such as Mingtien Coffee, Xeasetong Foot Massage, and Yongle Theaters join the league. A separate annex building is built for the first ever platinum five-star clubhouse in Ningbo. The clubhouse covers an area over 3,000 square meters, offering only 80 seats to the VIP members, which fully reveals its quality orientation.

Clear Water Luxury Garden International Water Club commits itself to be the largest recreational bathing plaza in East China, and with the innovative project positioning and fashionable design concept, it leads the brand-new industry standard.

碧水豪园国际水会，是斥资1.2亿、营业面积达30000平方米的超大型公共休闲项目。该项目除拥有超大型男女水疗区、600平方米室外游泳池、亲水观景露台之外，在RIG的建议下投资方于业界首次引入休闲Mall模式，名典咖啡、行易堂足疗、永乐院线等知名休闲品牌加盟，并单独设立宁波首座白金五星级会所附楼，会所部分面积逾3000平方米，仅设80席贵宾专位，足见其尊崇品质。

碧水豪园国际水会致力于打造华东地区规模最大的沐浴休闲广场，并以革新的项目定立、时尚的设计理念，引领全新的行业标准。

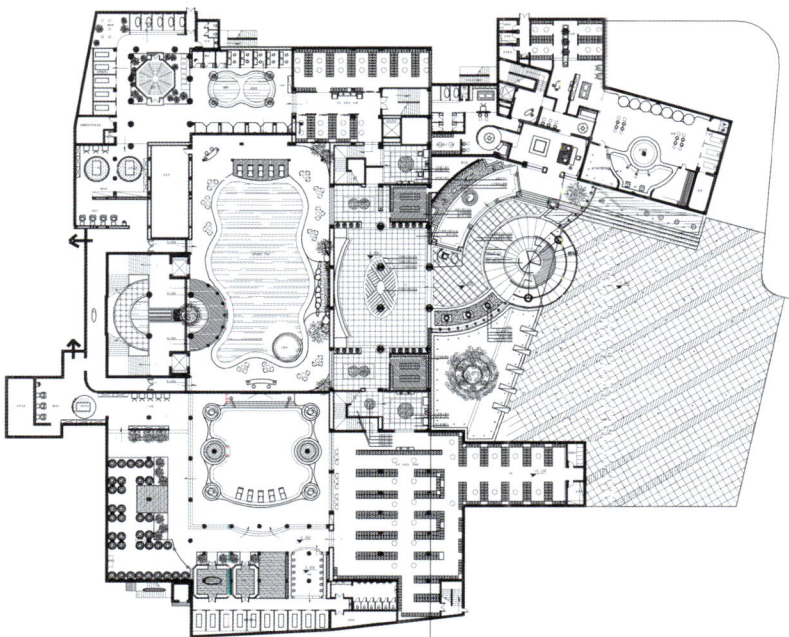

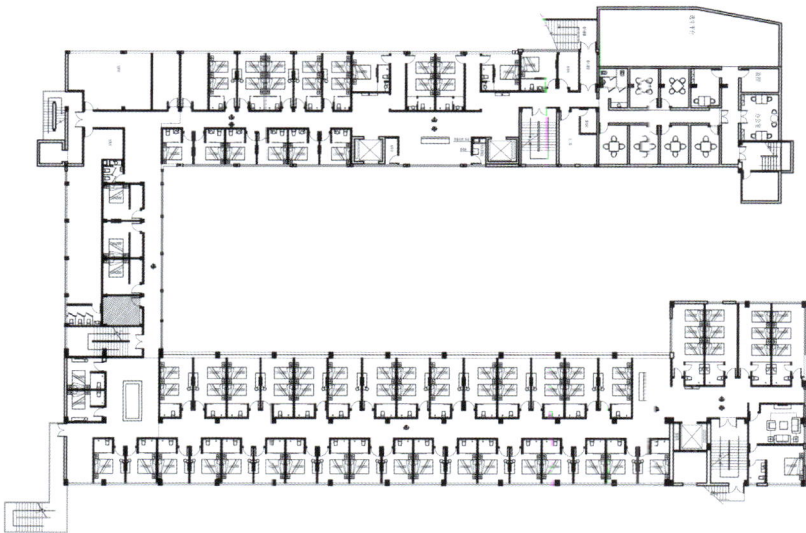
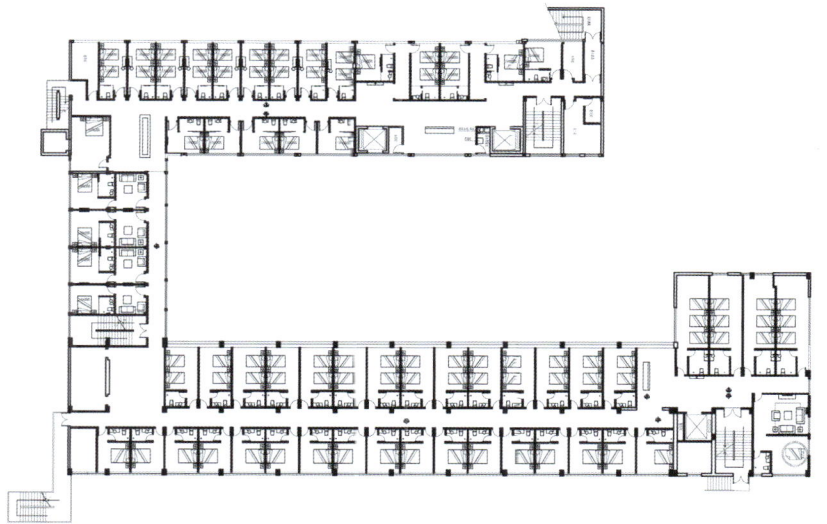

ZHONGLIAN · JIANGBIN ROYAL CLUBHOUSE

中联 · 江滨御景会所

— 地点

福建

— 面积

1800平方米

— 设计师

何海华、吴凤珍、高玲敏、吴华

— 方案审定

叶斌

— 设计公司

福建国广一叶建筑装饰设计工程有限公司

— 主要材料

木纹石、灰洞大理石、白色人造石、黑金沙大理石、福鼎黑火烧面、黑木纹石、15厘钢化磨砂玻璃、6厘大理石纹透光石、5厘茶镜玻璃、樱桃木面板

A PRIVATE FEAST
Entertaining Club
DESIGN

私享盛宴:娱乐会所设计

Zhonglian · Jiangyin Royal Clubhouse has exquisite details and reasonable arrangement, which, even in an unique way, presents a "dynamic" and "static" environment with elements of "sound" and "color".

Focusing on the theme of "sports, health, civilization, harmony and homage", the clubhouse meets the function of sales hall on the first floor. On basis of this, a round sand table display area is set up in the middle of the hall. Several geometric patterns which are made of wooden sticks constitute the main background and this makes the space more lively and penetrating. A negotiation area is set up on the right side of the hall. The designers crtfully adopt the method of cloister and by the use of tempered glass, three VIP reception areas are created which are both interesting and vivid. A bar counter is set up in the center of the second floor and on its left and right sides are the construction area and children recreation area respectively, with vivid and lively spatial route. The third floor is a book bar with large area of wooden decoration warming up the whole atmosphere. In here, you can enjoy a day of peaceful life while sitting at a table, with a cup of tea and a book at hand.

中联·江滨御景会所有精美的细部与合理的格局，更以独特的方式展示了"动'与"静"、"声"与"色"的环境。

该会所围绕"运动、健康、文明、和谐、尊崇"这一主题，在满足一层销售大厅的功能上，大厅中间设置了圆形的沙盘展示区，主背景采用木条构成几个几何图案，不仅生动，更使空间显得通透。大堂右侧设置了洽谈区、设计师巧妙地运用回廊的设计手法，用钢化清玻围合出三间VIP接待区，有趣且生动。二层的中间区域设置了吧台，左右两侧分别为建设区和小孩游乐区，空间路线灵动而活泼。三层为书吧，大量木饰面的运用使空间变得温暖起来，您可以享受一几、一著、一书、一天的宁静生活。

$$\frac{A}{S\text{-}01} \quad \frac{大厅\ A立面图}{比例\qquad 1:40}$$

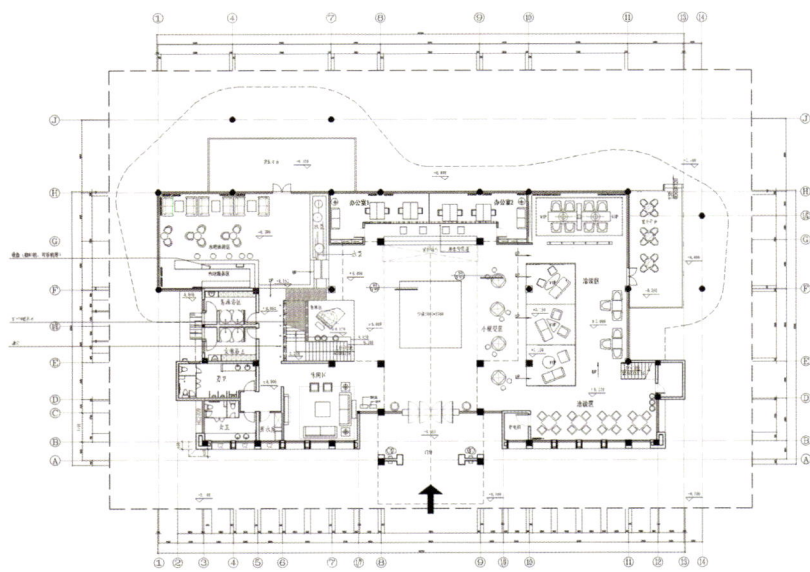

01
PM-01 一层平面布置图
比例　　1:100

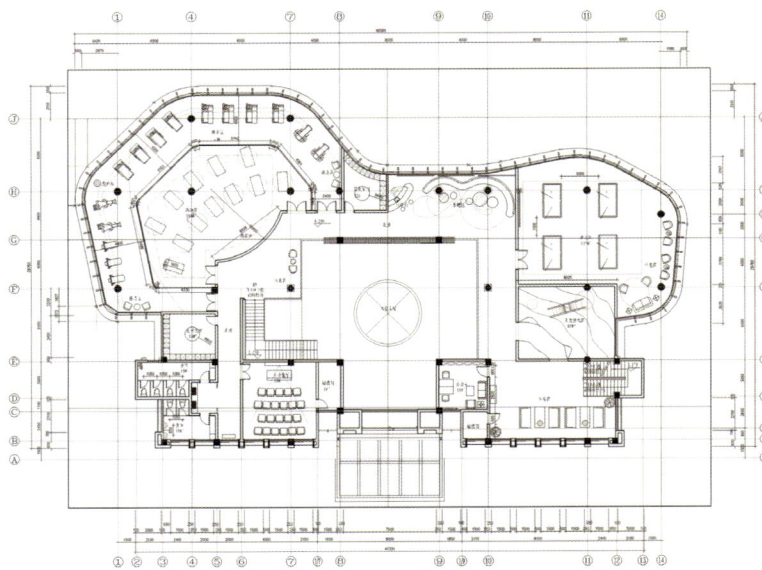

二层平面布置图
比例　　1:100

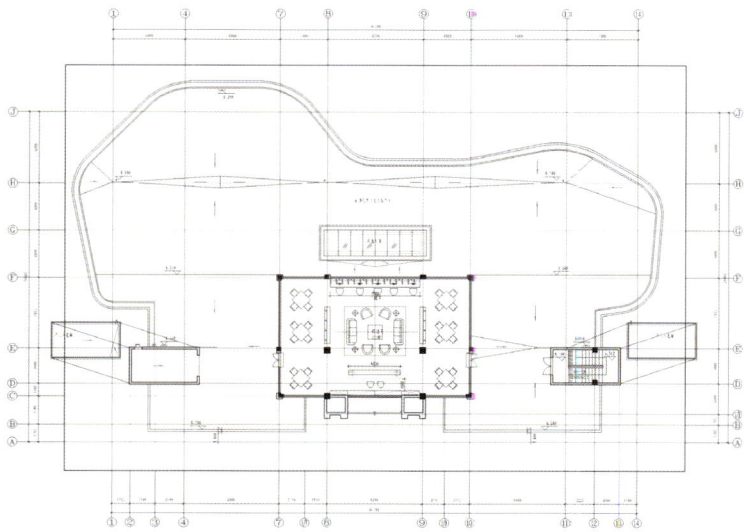

三层平面布置图

比例   1:100

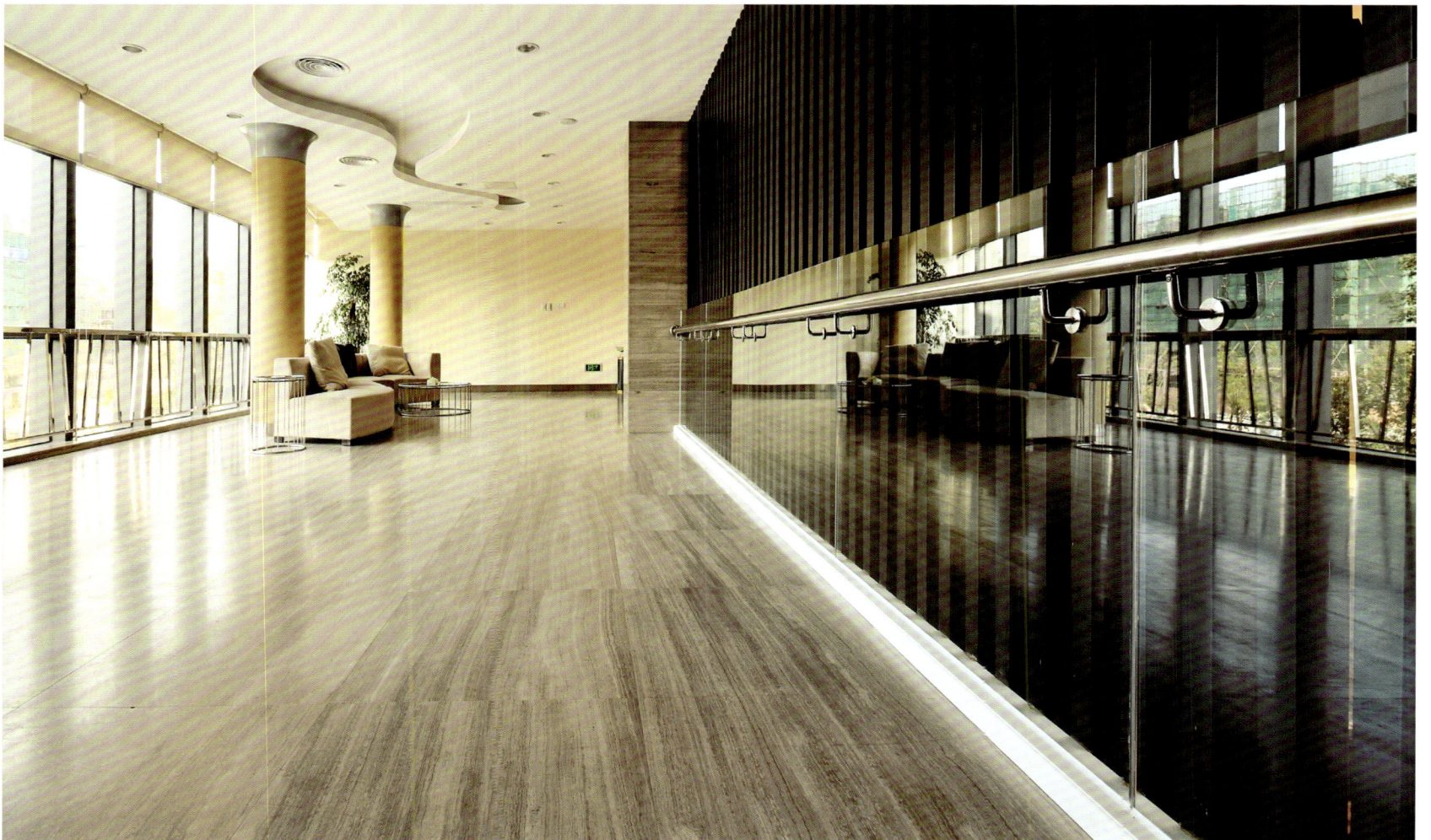

EYE CLUBHOUSE

眼神会所

—— 地点
佛山
—— 设计师
谢智明
—— 设计公司
大木明威社建筑设计有限公司（香港/广州/佛山）

A PRIVATE FEAST
Entertaining Club
DESIGN

私享盛宴：娱乐会所设计

Mosaic is a kind of inlay art originated from ancient times. It has experienced different historical backgrounds and has developed up to now with various materials and technics. Most people regard mosaic as a rich, colorful and diverse material for construction and decoration while ignoring its artistic essence. In this project, plenty of mosaic is used to create a space of art, a scene of leisure to promote thinking, thereby inspiring originality which gives people the desire to produce.

Pondering quietly, you will find mosaic both similar to the art of painting and sculpture and faculative in its artistic essence. In the Eye 2010 Originality Clubhouse, you will not only lay your eyes on a kind of decoration material with new technics but also might think further, about space, art, and life. In here, it offers a freer and more joyful lifestyle while at the same time finding a balance point among work, money and the quality of life to supply people with a sense of belonging.

马赛克是一种源于古代的镶嵌艺术，经历了不同的时代背景，以不同的材料工艺发展至今。大多数人认为，马赛克是一种丰富多彩、千变万化的建筑装饰材料，却忽略了其艺术本质。本案中，大量的马赛克创造出一个艺术空间、闲暇的场景，促人思考，从而激发创意，让人生出创作的欲望。

若你静静思考，便会发觉马赛克既与绘画、雕刻艺术有相同之处，又有它的偶发性的艺术本质。在眼神2010创意会所中，你的眼睛看到的不单是一种新工艺的装饰材料，或许你会有更深的思考，对空间、对艺术、对生活。在这里，它能为你提供更自由、更愉悦的生活方式，同时在工作、金钱和生活质量之间找到一个平衡点，让人找到归属感。

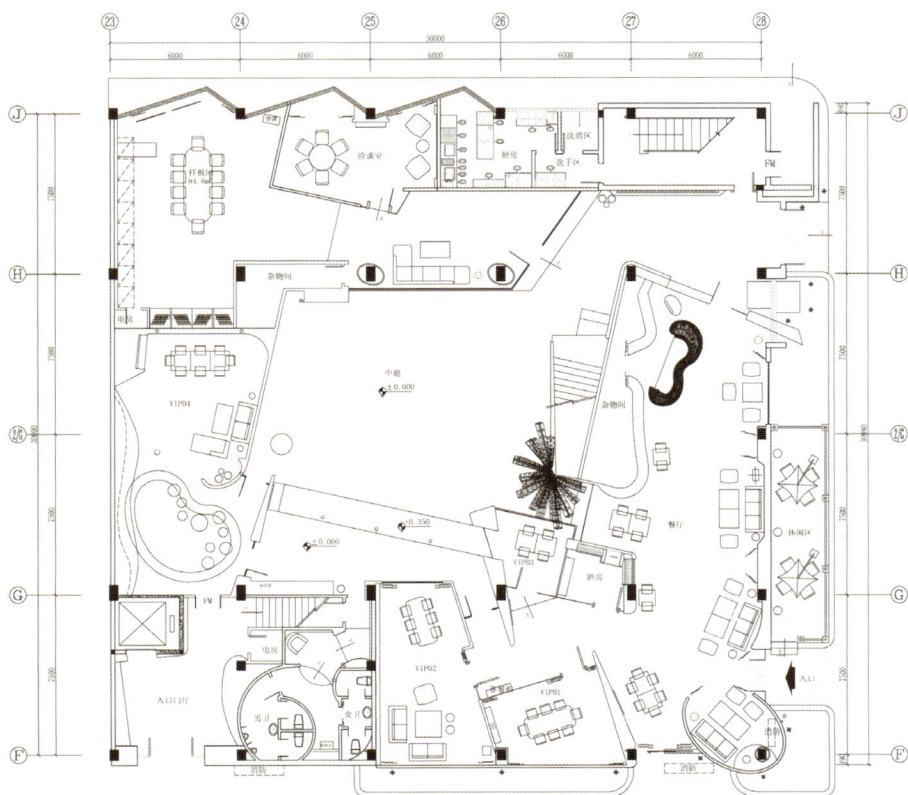

JINGQISHEN HEALTH CARE HOUSE

精气神健康养生馆

—— 地点
东莞

—— 面积
2000平方米

—— 设计师
符军

—— 设计公司
RYB · 三原色建筑装饰设计院

—— 主要材料
灰镜、柚木木饰面、仿柚木金刚板、透光片、刨
花板、爵士白大理石、布艺软包、艺术马赛克、
地胶、条形混拼地砖、线帘

A PRIVATE FEAST
Entertaining Club
DESIGN

私享盛宴：娱乐会所设计

The designer prefers to draw inspiration from the structure of prominent modeling. He is enlightened by a little, light leaf and composes a melody suitable for the space, clear and poppling. Green is a color of nature, and Jingqishen Health Care House is a business space with its major services like heath preserving, physical therapies and health care. The introduction of green leaf exemplifies the theme and charm of the space exactly.

The elegant design style on the whole, the simple and graceful modeling of the house, the dense and mature colors, all these present the space hierarchy and plain materials. Not only does the wall decoration with wooden texture make up for the monotonous decoration of walls, but the most important thing is that it provides a contrast with the green leaves on the ceiling, and coheres with each other as well, which adds romantic feelings to the leisure atmosphere. The light plays a role of magician in the setting and unifies different design elements, different surroundings and different people.

设计师喜欢从造型突出的结构中寻找灵感。他从一缕轻叶中得到启发，进而谱写出迁于空间的乐曲、清晰、荡漾。绿色是属于自然的颜色，"精气神"的特性是以养生、理疗、健康保健为主要经营项目的商业空间，绿叶的引用恰好体现出空间的主旨和韵意。

设计整体格调高雅，造型简朴而优美、色彩浓重而成熟，宣扬空间的层次感与朴实的材质感。木质线条的墙面装饰不仅弥补了墙面装饰的单调，最重要的是与天花上的绿叶形成对比，相互协调，为休闲的气氛增添了一丝浪漫情怀。光线在环境中扮演着魔术师的角色，统一空间中不同的设计元素、不同的环境、不同的人。

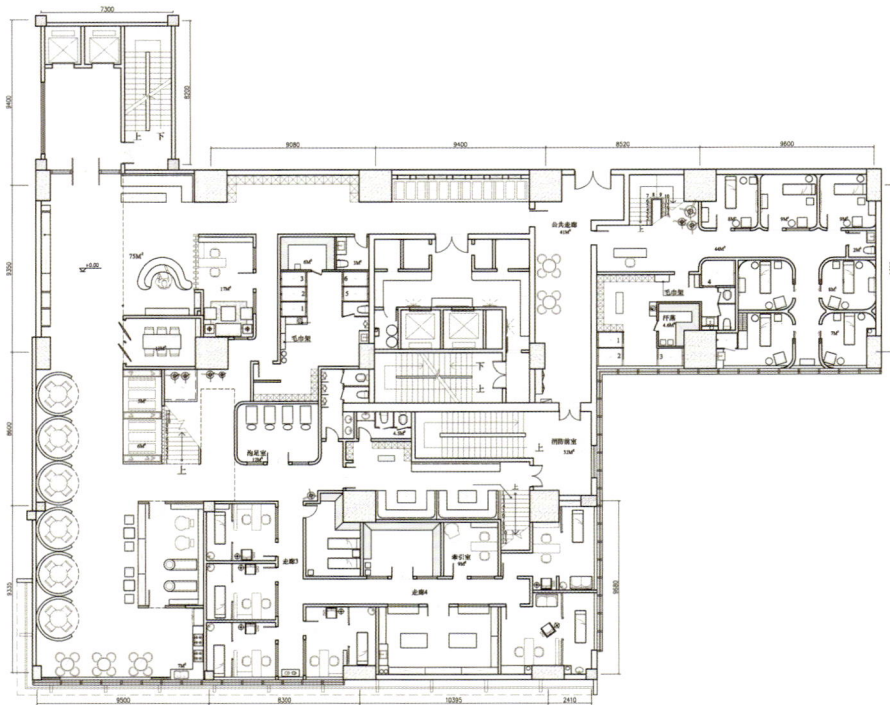

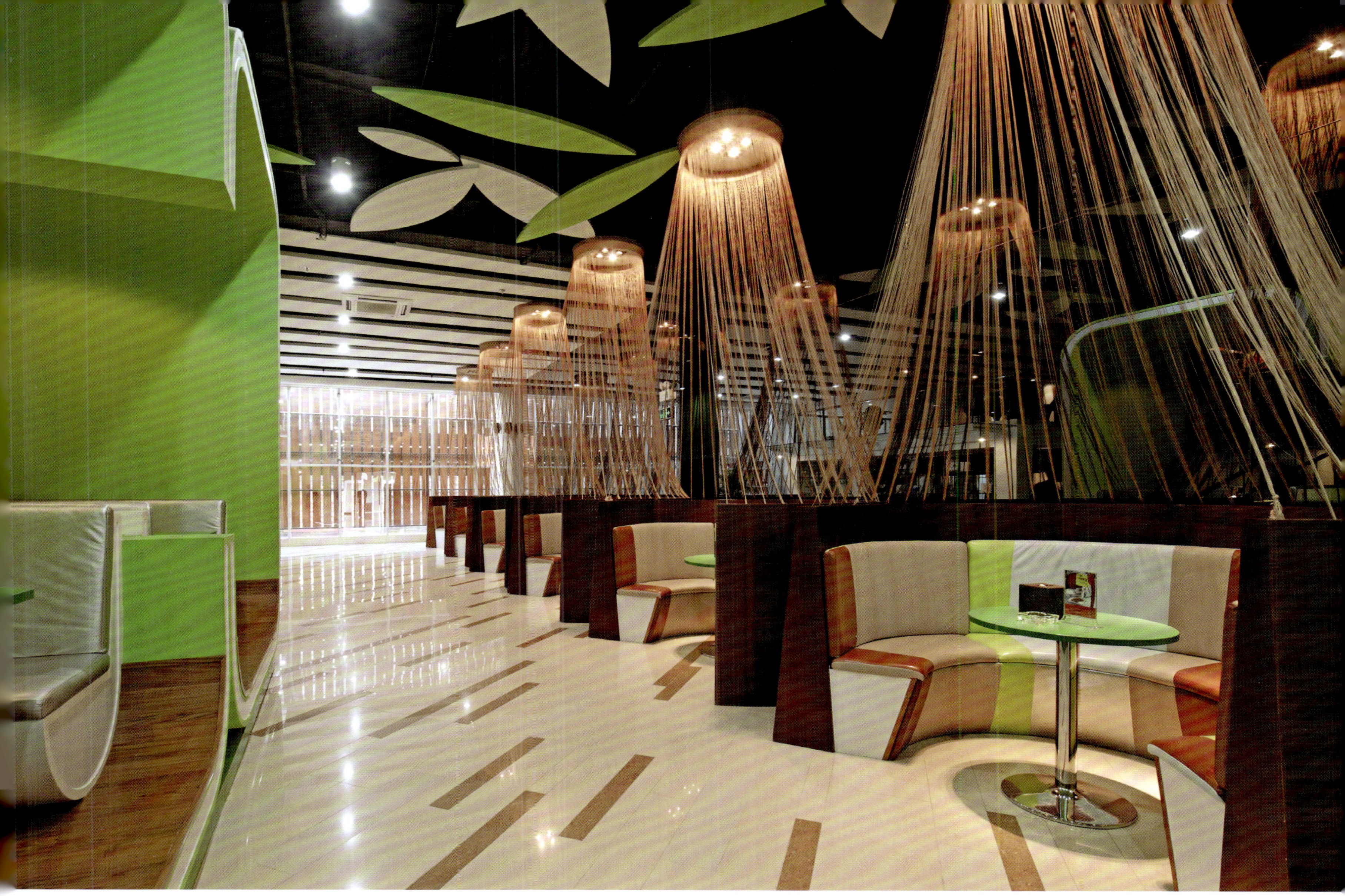

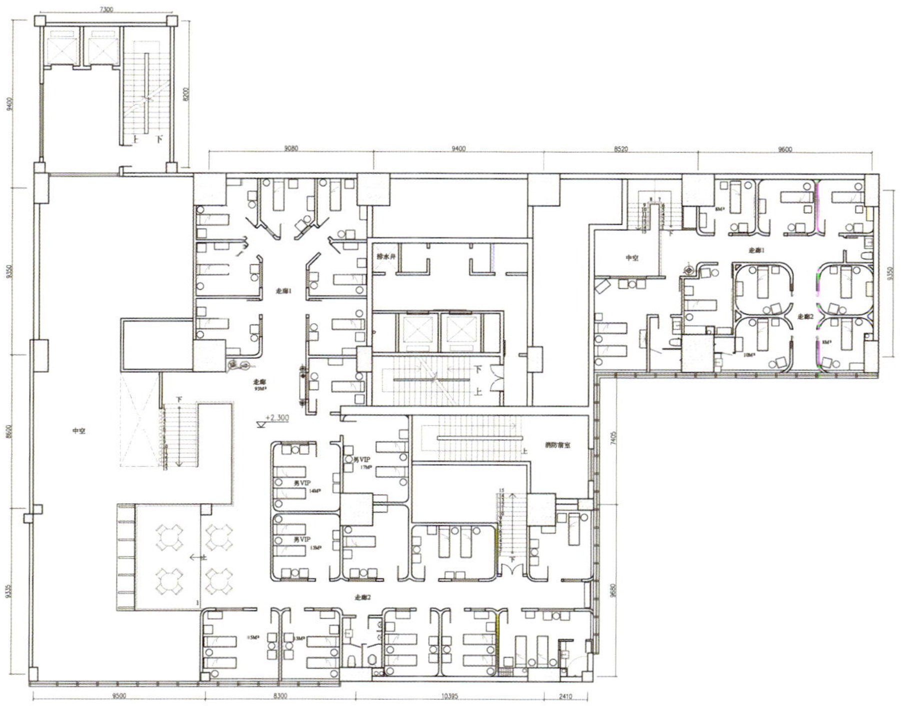
7300
9080
9400
8520
9600
中空
9500
8300
10395
2410

JIULE CLUBHOUSE, HEFEI

合肥九乐会所

— 地点
合肥
— 面积
1000平方米
— 设计师
施旭东
— 设计公司
Sunever 设计　旭日东升装饰机构
— 主要材料
镜面拉丝黑钛金、黑金龙大理石、皮革编织装
饰、蘑菇面芝麻灰花岗岩、黑镜、银色仿古镂空
画、壁纸、擦色欧饰墙板

A PRIVATE FEAST
Entertaining Club
DESIGN

私享盛宴：娱乐会所设计

This project draws design inspiration from traditional oriental elements, which are bravely destroyed and denied; thereby it creates a modern conception. Chinese traditional lacquer painting and bright Chinese lacquer are full, deep and mottled with a sense of texture understated. The design of the traditional red stoneware pot with loop handle was bravely destroyed with its changes of design styles developing into a functional reception desk, which, meanwhile, resembles the pure-white work of art in the lobby; pearls baptized with glass, light and shadow draw a beautiful curve in order, with flickering candlelight, candlesticks; simple and crude stone hitching posts; metal accessories and blue and white porcelain patterns. The designer employs the tension produced by the interweavement, contrast and harmony of colors, materials, light and shadow and modeling to arouse the resonance of the visitors and convey the inner oriental spirit of the space.

本案从东方传统元素中汲取设计灵感，大胆地加以"破坏"和"否定"，从而创造出一个具有时代气息的意念。中国传统磨漆画艺术、浓烈的红色大漆、饱满深沉而斑驳、肌理感含蓄内敛。传统紫砂工艺的提梁壶被大胆"破坏"，设计形式演变成具有功能性的接待前台，同时又如大厅中的纯白色艺术品；一颗颗珍珠透过玻璃和光影的洗礼，有序地拉出优美的弧线；摇曳的烛光、烛台；朴拙的石材栓马桩；金属装饰品以及青花瓷的图案。设计师利用色彩、材质、光影以及造型的穿插、对比、和谐所产生的张力，来引起来者的共鸣，传达出空间的东方内在精神。

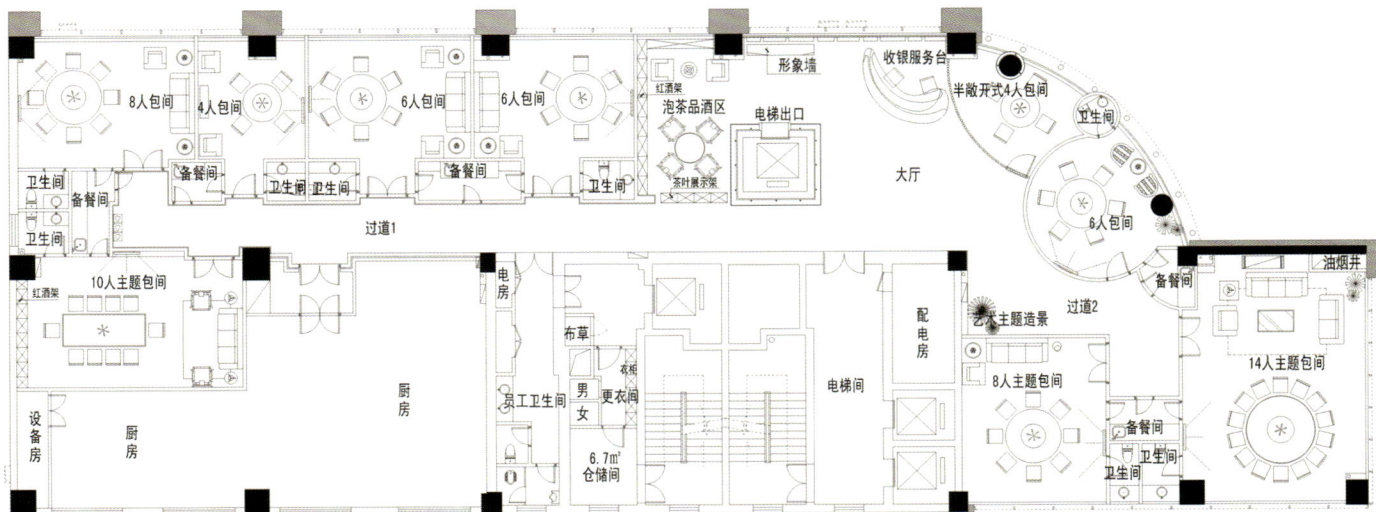

形象墙
收银服务台
半敞开式4人包间
泡茶品酒区
红酒架
电梯出口
卫生间
8人包间
4人包间
6人包间
6人包间
茶叶展示架
大厅
6人包间
卫生间
备餐间
卫生间
卫生间
备餐间
卫生间
过道1

红酒架
10人主题包间
电房
配电房
艺术主题造景
过道2
油烟井
布草
衣帽
员工卫生间
男女更衣间
电梯间
8人主题包间
14人主题包间
设备房
厨房
厨房
6.7㎡仓储间
备餐间
卫生间
卫生间

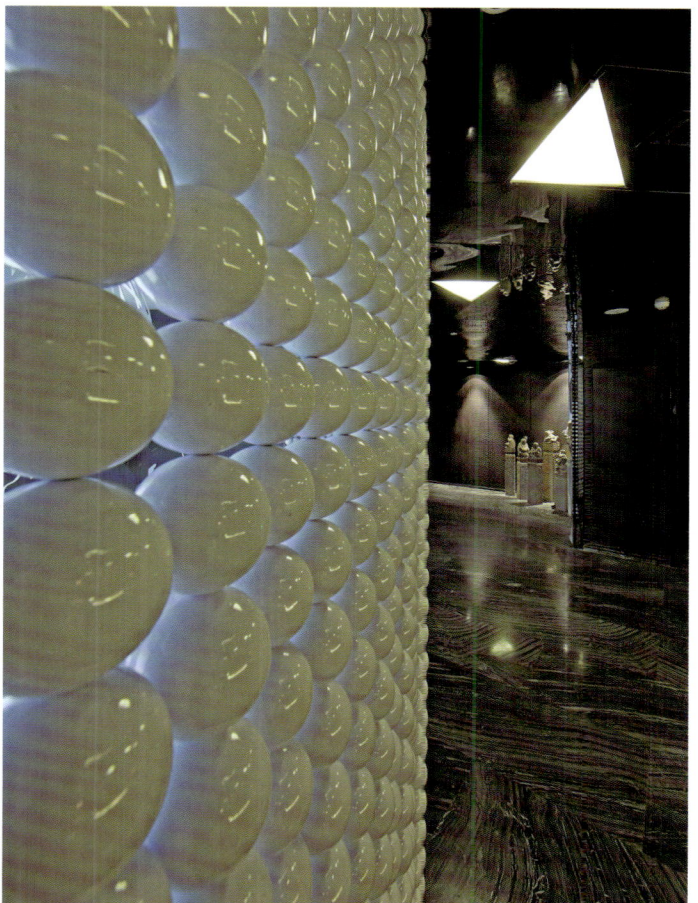

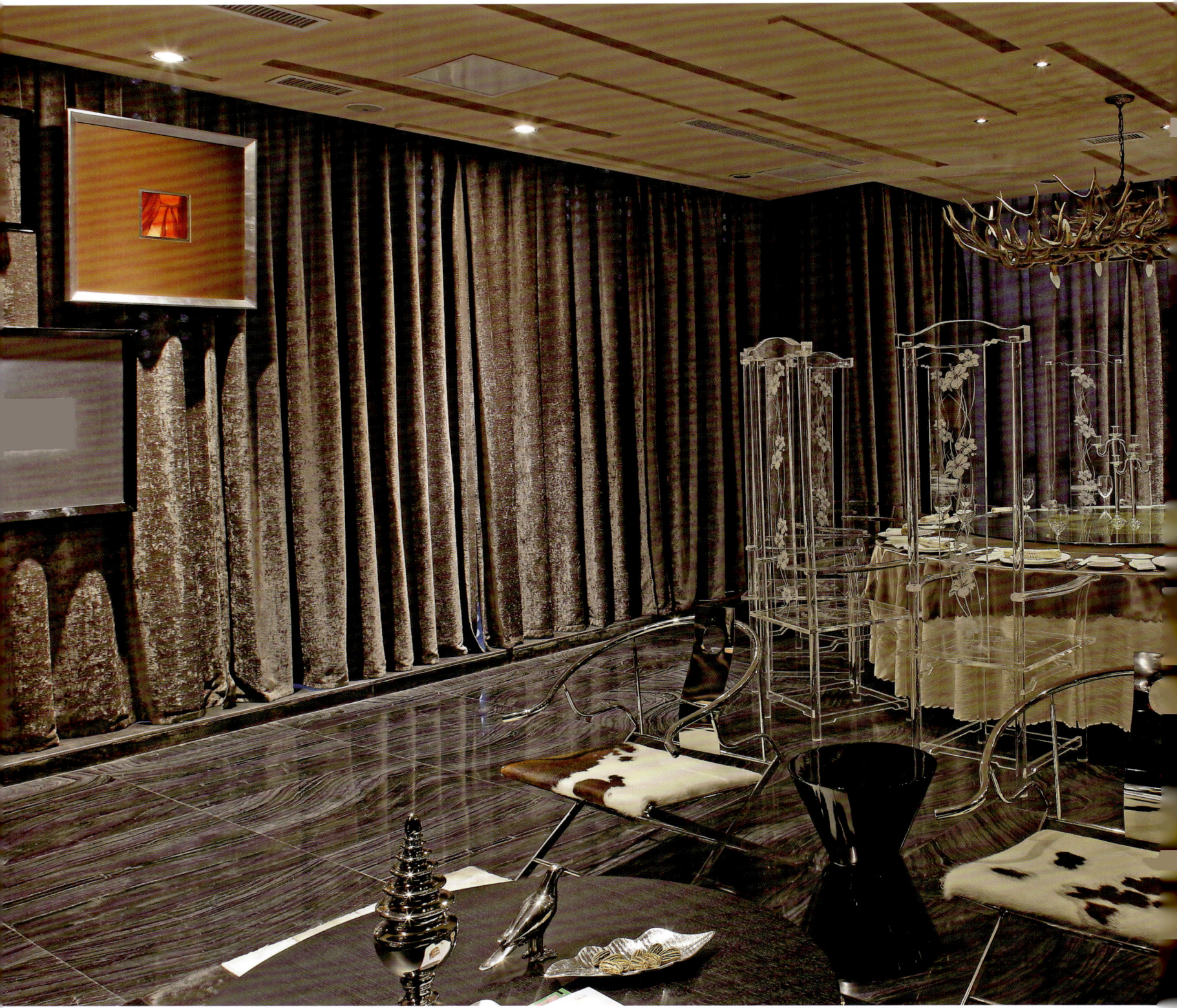

NINGBO VANKE GOLDEN SEASIDE CLUBHOUSE

宁波万科金色水岸会所

— 地点

宁波

— 设计公司

上海乐尚装饰设计工程有限公司

LESTYLE
樂尚設計
WWW.LESTYLE.COM.CN

乐尚设计成立于1999年，是一家从事高档楼盘
会所、售楼处、样板房、别墅、软装配饰与施
工于一体的专业室内设计公司。乐尚人以"创
意自由、规划严谨"的理念，将创意和规划优
势进行整合，最大限度地为客户创造价值。

A PRIVATE FEAST
Entertaining Club
DESIGN

私享盛宴；娱乐会所设计

The interior design of this project is based on the principle that 'letting construction applaud life and letting materials naturally present'. It adopts the style of mix-and-match, and makes materials undecorated to show their original looks. The free match seems simple, but reveals the profound design capability of the designer.

At the reception area, the red tiles remain naturally uncovered, and the rugged design forms the vision of a cylinder, which displays the original wildness as well. The axis-symmetric design is adopted by the two sides, the wooden display cabinet on one side and the steamy window on the other side form the contrast between illusion and reality. The design of the ceiling mainly employs the dark tone to match the design of the entire space, and it also responds to the aging treatment of the floor. The design of misaligned sham beams displays the unparalleled quality of the space, and weakens the feeling of oppression caused by the disadvantage of the ceiling height.

本案的室内设计以"让建筑赞美生命、让材质自然呈现"为理念出发，风格上采用混搭，材质上让其裸露，以展现原始材质的风貌。看似简单的自由搭配，却彰显着设计者深厚的设计功底。

前台接待区，红色的墙砖自然裸露，凹凸不平的设计形成了柱体的视觉，也有了原始的粗犷。两边的设计呈轴对称，一边的木架展示柜与另一边的雾面玻璃形成虚实的对比。天花的设计主要采用暗调来搭配整个空间的设计，也呼应了地面的做旧处理手法。槟错假梁的设计展现出空间无可比拟的气质，淡化了层高不足给人的压抑感。

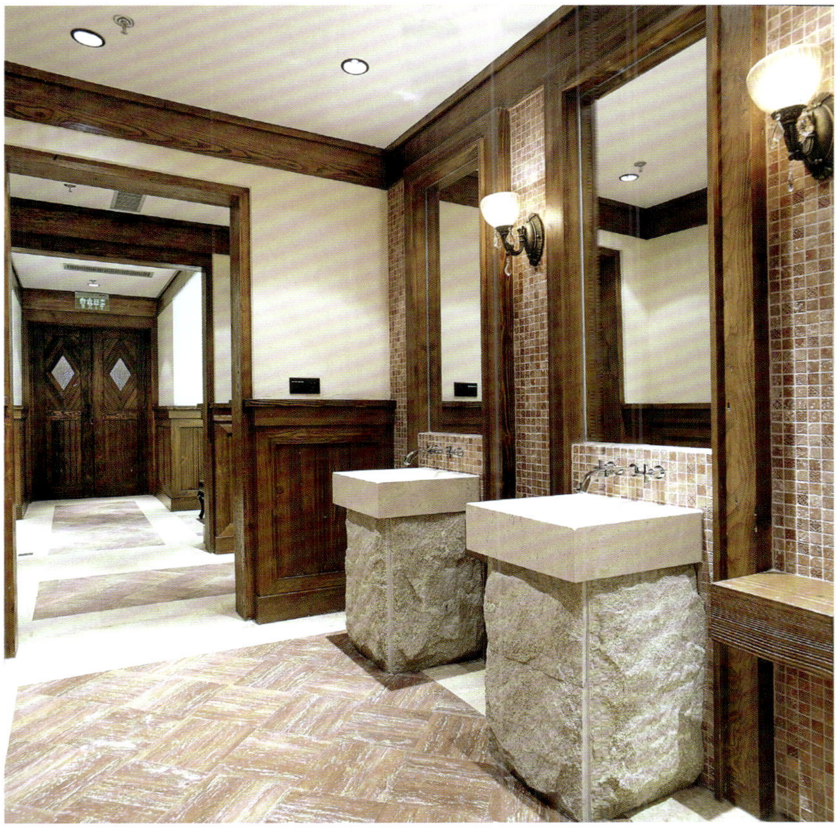
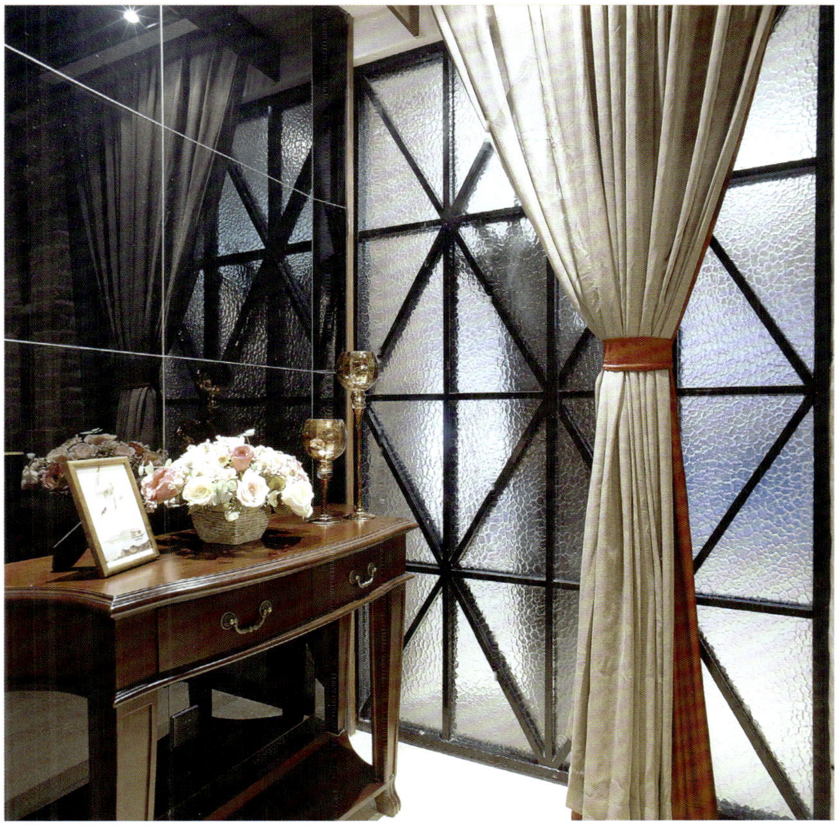
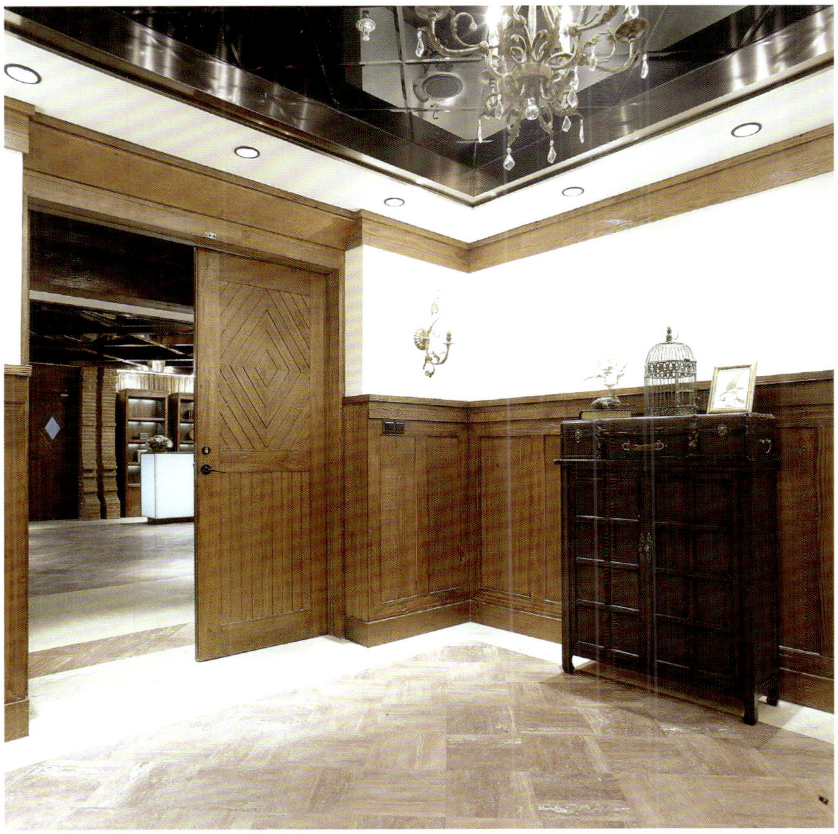
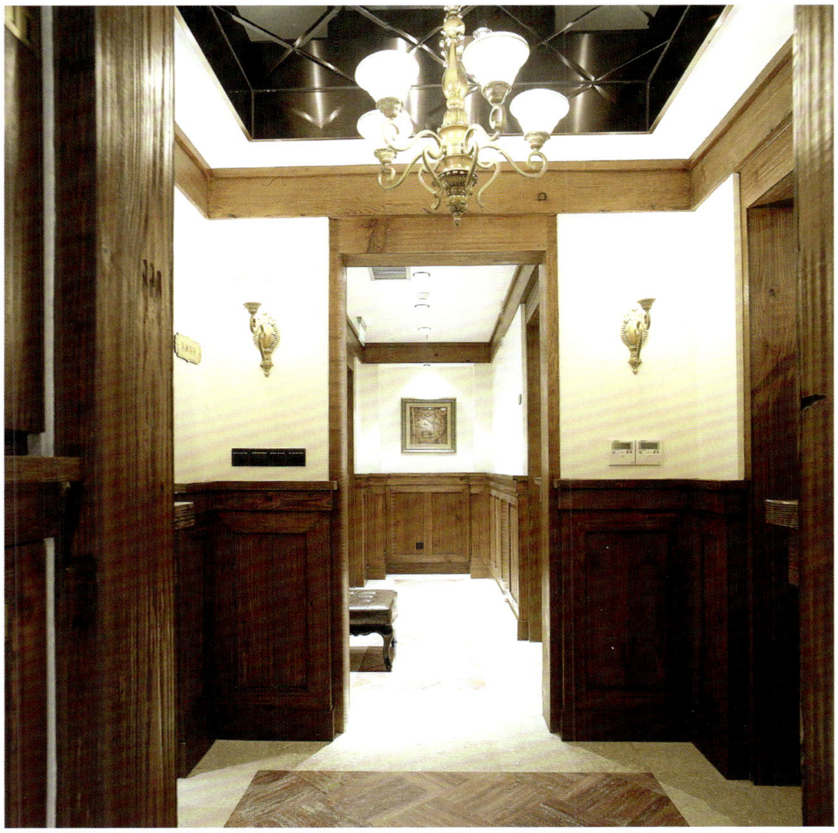

BROWN SUGAR

BROWN SUGAR

—— 地点
中国台湾台北

—— 面积
室内198平方米、室外549平方米

—— 设计师
甘泰来

—— 设计公司
齐物设计事业有限公司

—— 主要材料
和平白石材、橡木染深色、镜面不锈钢、灰色洗石子

A PRIVATE FEAST
Entertaining Club
DESIGN

私享盛宴；娱乐会所设计

Brown Sugar in Taipei follows the design of Shanghai branch which adopts the concept of amphitheater. Sections of different heights spread in the open area and they are the bar counter section, sofa section, terrace compartment section, table section, stage section and VIP section. In the front of the base is the elevated bar counter section, and the stage section is on the opposite, and the scattered table section is in between. The left and right sides are the elevated sofa section and tiered seats.

The deliberately widened aisle in sofa section and the arrangement of the furniture being away from the walls leave a lot more freedom to the motion lines. The vein-like patterned screen, the vertical surfaces and the platforms are all decorated with the stainless steel frame and dark color mirrors, and dark-brown glass are massively used to create a superb vision like neon lights.

Brown Sugar台北店在设计上承袭了上海店"阶梯剧场"的概念，在开放空间内规划多种平台高度，共划分为吧台区、沙发区、阶梯包厢区、散桌区、舞台区、VIP区。基地前端为架高吧台区，与尽头舞台区中包夹着散桌区，左右两侧为架高的沙发区与阶梯式的座位。

刻意放宽的沙发区走道与家具不贴墙的摆设方式，使得动线更为自由。以不锈钢材质作为人字纹屏风以及立面、地台的框饰，并大量使用深色镜面及茶色玻璃，使空间中镜射出霓虹光般的华丽幻影。

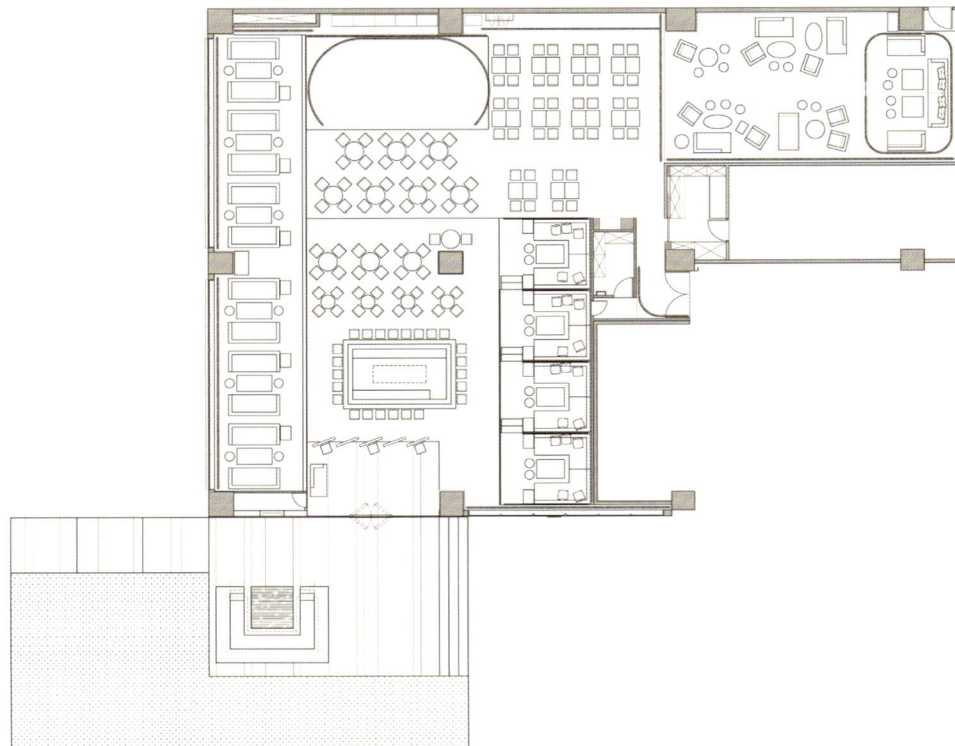

WUHAN WORLD TRADE CINEPLEX BAMBOO FUN FOOT BATH

武汉世贸影城竹乐浴足

—— 地点
武汉
—— 面积
1000平方米
—— 设计师
韦文生
—— 设计公司
广州文智装饰设计工程有限公司

A PRIVATE FEAST
Entertaining Club
DESIGN

私享盛宴，娱乐会所设计

This project is designed to promote a healthy and peaceful attitude to life. Moreover, in an eco-friendly and recreational environment, it attempts to create such an atmosphere in which people can appreciate health and enjoy life from daily lives, and adopt positive attitude of happiness and contentment. At the entrance, the lovely foot-shaped sign uses green color and the bamboo-leaf lamp cord to highlight the theme of the space — bamboo, brightness and liveliness. In the indoor space, the log, tawny mirrors, and the nostalgic color of the floor coverings, display not only the fashion taste, but a touch of natural flavor as well. They also enable people to feel high profile, elegancy as well as intimacy without the sense of distance. With such non-distanced interaction, people can feel and enjoy life from the bottom of their hearts.

本案旨在提倡一种健康、安乐的生活状态，并力图用一个环保、康乐的环境为人们创造这样一种氛围：让人从平日的生活中感受健康、享受生活，同时培养知足常乐的健康心态。入口处可爱的脚掌标志运用绿色和竹叶灯带表现出空间主题——竹、鲜明、活泼。室内空间里，原木材料、茶镜、怀旧色的地面铺装，既显出时尚品位，又流露出淡淡的自然气息；让人感受高调、优雅的同时也感觉亲近，不会有距离感。有了无距离的接触，才能从心底感受生活、享受生活。

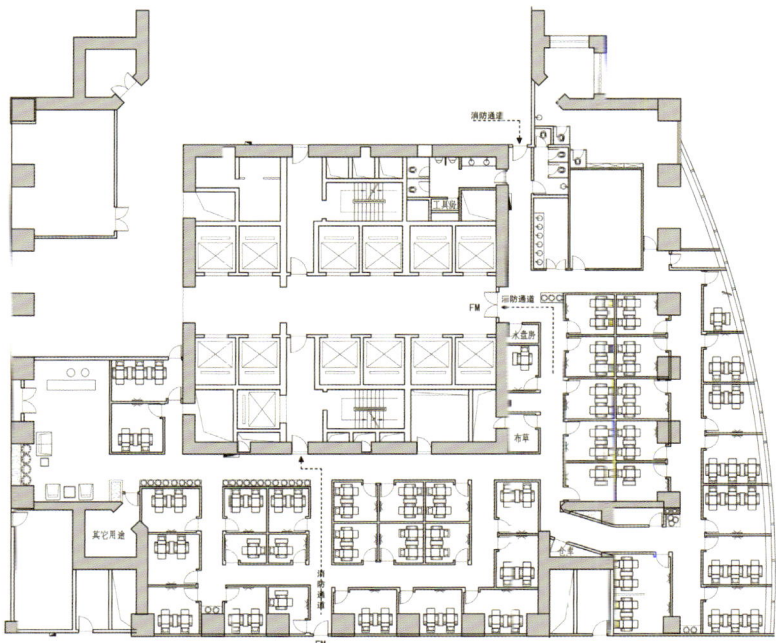

AHMANSON FOUNDERS ROOM

阿曼森创办室

— 地点
洛杉矶
— 面积
232平方米
— 设计师
Hagy Belzberg
— 摄影
Fotoworks

A PRIVATE FEAST
Entertaining Club
DESIGN

私享盛宴: 娱乐会所设计

The Ahmanson Founders Room is a 232m² addition buried in the first level of subterranean parking at The Music Center in downtown Los Angeles. The sunken location of the room coupled with an almost clandestine preoccupation of exclusivity by the founders helped orient the design objectives. Belzberg Architects pursued the development of sensual lighting schemes and unique applications of material and texture to create a warm place of respite between the congested streets of L.A. and the brimming communal areas of The Music Center on event night.

The materials used for the ceiling and walls were actually low-grade douglas fir lumber and medium density fiberboard. However, sculpting the ceiling panels and perforating the wall panels with supple patterns and textures transformed low-grade into classy with stunning visual effects and a colorful saturation of the space. The character and unique components of this room have become an entity for which the Ahmanson Theatre founders can identify with and claim as their own.

阿曼森创办室是洛杉矶商业区音乐中心的一个附属建筑，占地232平方米，建筑物底部嵌入到音乐中心的地下一层停车场之中。凹陷的地理位置，加上创办人对于近乎保密的独享性的考虑，为设计目标明确了方向。贝尔兹伯格建筑设计事务所致力于开发一些感官明亮的设计方案并运用特有的装饰材质，在熙攘的洛杉矶街头和音乐中心盛典之夜拥塞的公共场所之间营造出一个温馨的休憩场所。

天花板和墙壁的用材实际上只是低等的花旗松木料和中密度纤维板。然而，雕花的顶棚镶板以及带有灵动图案和质感的镂空墙板化低栏为上乘，带来震撼的视觉效果和空间上的高色彩饱和度。创办室鲜明的特色以及特有的包厢已经成为阿曼森剧场创办人进行身份识别和主张专属权的有效实体。

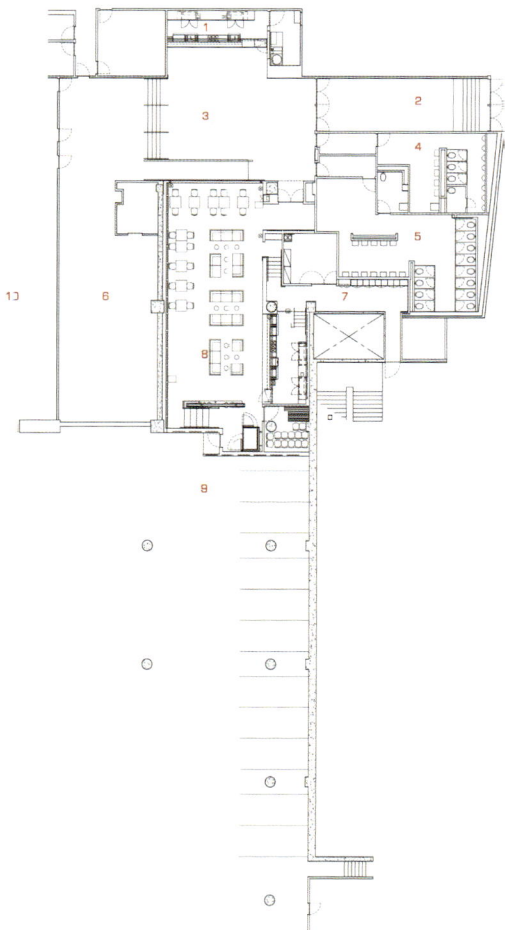

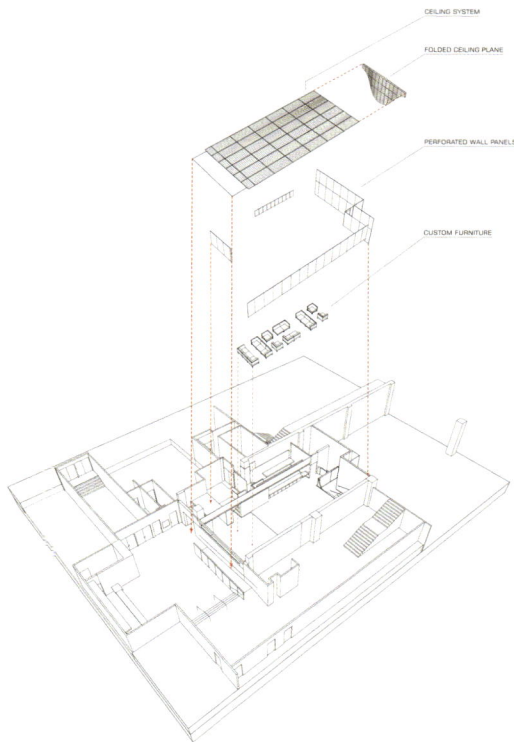

CEILING SYSTEM

FOLDED CEILING PLANE

PERFORATED WALL PANELS

CUSTOM FURNITURE

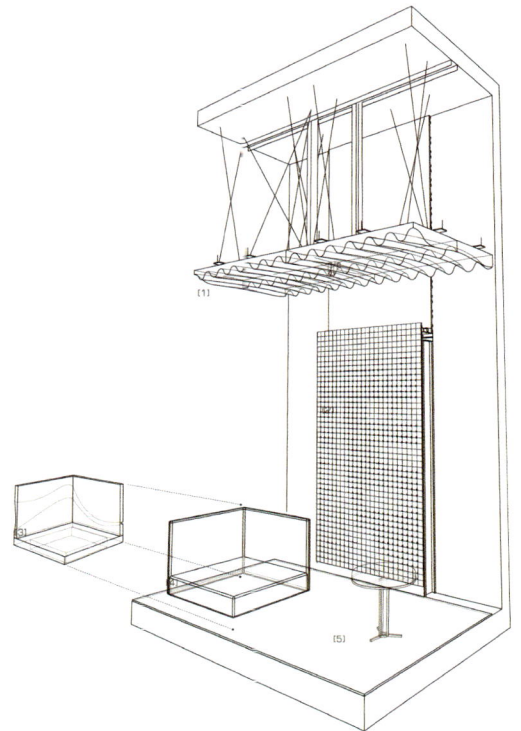

[1]

[5]

Facing West toward garge entry

ENJOYING THE GRAND BANQUET

私享的盛宴

—— 地点
长沙

—— 面积
122平方米

—— 设计师
陈志斌

—— 参与设计
李智勇、谭丽、彭辉

—— 设计公司
鸿扬集团陈志斌设计事务所

—— 主要材料
灰镜、罗马洞石、银箔、地毯

—— 摄影师
周重山

A PRIVATE FEAST
Entertaining Club
DESIGN

私享盛宴：娱乐会所设计

No hustle and bustle, only elegant bearing bloomed in quietness.
No contradictions, only harmonious dialogue between the Chinese and the western styles.
In this low-keyed yet gorgeous space, you will be able to search for the true meaning of aesthetics with customers who have got excellent tastes, advanced education and high appeal. The creative idea is like the floating cloud or the flowing water, neither of which is predictable but can easily hide or emerge in the space. The furniture appears slightly jumpy in the elegant atmosphere, with the dignified pieces sitting respectfully while the obliquely ones placed carrying distinguished air of grace. The design of one area is rigid and straight and that of the other is curving and soft, which makes the former a precise and solemn space, appropriate for receiving a new guest and the latter a relaxing and refreshing one, suitable for communicating with an old friend. Both of them carry distinctive characteristics and play with Art Deco Spirit. All the lattices and corbel arcch with Oriental elements are painted in an elegant pearl color and become shining devices. When the east meets the west, after the collision of passions, the perfect blending, even without the make-up will be able to merge into the trends of today's designing styles.

没有喧嚣，典雅气质沉静绽放。
没有矛盾，中西时尚和谐对话。
在这个低调、华美的空间里与高品位、高学历、高诉求的客户一同寻找美学的真谛。创意如行云流水一般，云水无定式，在空间中自由地若隐若现。典雅氛围中，家具略显跳跃，端庄的正襟危坐，斜摆的仪态万方，一刚一柔，一曲一直，一个适合初次会客，严谨庄重；一个适合熟客交流，轻松休闲。两大会客区特色鲜明，都以Art Deco的精神游戏其间，东方元素的花格、斗拱都披上了典雅的珍珠颜色，成为耀眼的装置。当东方遇见西方，激情碰撞之后，褪尽铅华，终将融汇于当代创意的潮流中。

鸿扬家装优仕设计私享

Hiyun

YALLANS WINE CLUB

亚兰尼斯红酒会所

— 地点
南京

— 面积
300平方米

— 设计师
李浩澜

— 设计公司
南京浩澜设计事务所

— 主要材料
白石材、白纱帘

A PRIVATE FEAST
Entertaining Club
DESIGN

私享盛宴：娱乐会所设计

This is a collision of the East and the West. The architecture is of Chinese style, while the red wine reflects western style, so the biggest challenge for the designer is to maintain the architecture's eastern features and the club-house's red wine atmosphere at the same time. Finally the designer determines to try a different Chinese style, which focuses only on the artistic conception. Interpreting Chinese style as a pure artistic conception, the designer starts with white color, adopts some elements of Chinese neoclassical interior design style, and utilizes modern techniques, even some western elements, to express the pursuit of the oriental spiritual state: elegant and modest, dignified and magnificent. In addition, the designer integrates the essential features of a red wine club into the design to make an attempt to unite the east and the west, which results in an elaborate production from the team, although not a perfect one.

这是一次东西方的碰撞！建筑是中式的，红酒是西式的，如何保持建筑的风貌和红酒会所的感觉成为设计的最大难点。最终设计师索性做出一个不一样的中式，只取意境。将中国风格理解为纯净的意境，从白色调入手，通过中国新古典室内风格的元素，用现代的手法甚至是西式的元素表达对清雅含蓄、端庄丰华的东方精神境界的追求。同时将红酒会所应具有的一些特质融入其中，做了一次中西合并的尝试，结果虽不能尽善尽美，却是设计团队用心付出的成果。

JINZUN CLUB

金尊俱乐部

— 地点
广州

— 面积
2787平方米

— 设计师
曾卓中、葛新亮、袁润光

— 设计公司
广州市品祺设计有限公司

A PRIVATE FEAST
Entertaining Club
DESIGN

私享盛宴：娱乐会所设计

The design is the reflection of the planning idea and the project is positioned to be noble yet fashionable as well. The designers start the project from the aspects of modeling, hues and lighting, which brings pleasant surprises of vision, senses and psychology. In modeling, the ancient Greek majestic columns are regarded as the fountain of inspiration, extracted and recombined by the standard of post-modern styles, which expresses the magnificence of nobility. In the aspect of hues, the basic color looks calm and steady and the club is mainly embellished with gold and silver color and decorated with enchanting soft furnishings, which implies beauty and softness in nobility and magnificence. In lighting, the design pursues the thorough display of nightclub's noble and fashionable atmosphere in the light and shadow with multi-elements.

设计是策划思想的体现，本案定位高贵而不失时尚。设计师从造型、色调、灯光这三个方面作为切入点，带给人视觉、感官及心理方面的三种惊喜体验。造型上，以古希腊宏伟的柱廊作为设计灵感的源泉，以后现代的感觉加以提取与重组，表达出大气的高贵。色调方面以沉稳为基调，用金、银作为重心的点缀，并略配媚态的软装配衬，使得高贵、大气中暗含娇柔。在灯光方面，力求通过多元的手法使夜总会高贵时尚的气氛在交错的光影中体现得淋漓尽致。

金爵19

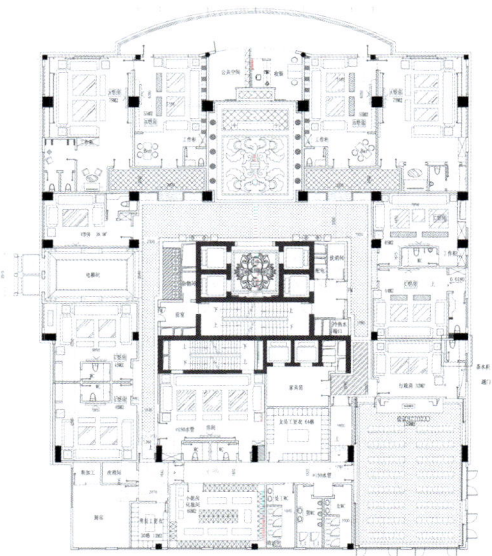

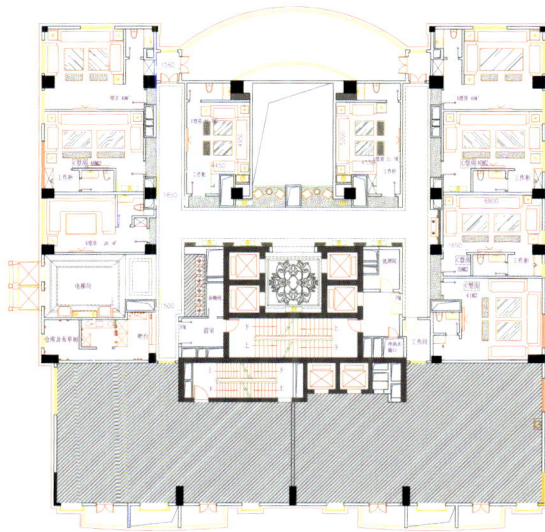

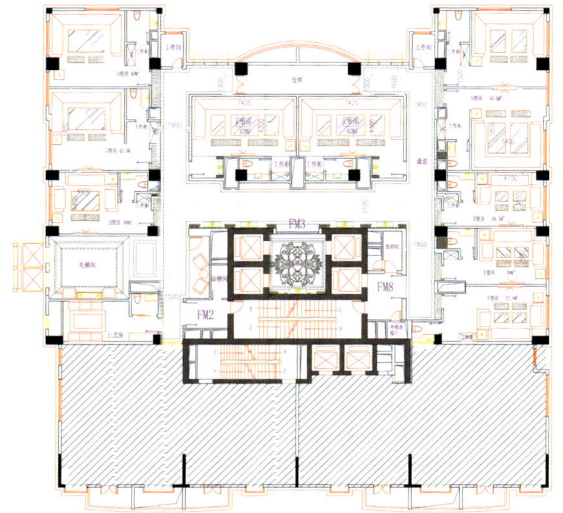

BEIJING XIANGSHUI TEAHOUSE

北京湘水茶园

—— 地点

北京

—— 面积

380平方米

—— 设计师

鞠千秋

—— 主要材料

背漆玻璃、壁纸、仿古砖、实木、汉白玉

A PRIVATE FEAST
Entertaining Club
DESIGN

私享盛宴，娱乐会所设计

This project is designed to give the impression of an old Beijing scene recalled from memory and to bring people the sense of neighborhood intimacy. Entering the teahouse, you seem to step into the Beijing hutong. The melodious bird singing and a slight aroma of the tea spread over every corner of the space. Wandering in the corridor, you may even neglect the fact that you are in an interior space. There are bridges, water, hutong and the courtyard, which all together form an epitome of the old Beijing scene. With the blue bricks, grey tiles and the log in view, breathing the natural air, you will feel the intimacy and comfort of nature. An artistic image of small bridge over the flowing water is created by the reproductions of antique bricks and the white marble. Together with the traditional wallpaper and wood structures, the lingering charm of Chinese flavor is naturally revealed, just as they were used to be.

本案设计力图打造一个印象中的老北京场景，带给人邻家别院的亲切感。走进茶馆的瞬间，就仿佛步入了北京的胡同，悦耳的鸟鸣、淡淡的茶香弥漫在空间的角角落落。穿梭在走廊里，甚至让人忽略了这是室内空间，这里有桥、有水、有胡同、有院落，这是一个老北京场景的缩影。入目的青砖灰瓦和原木材料，呼吸着自然的空气，让人感受自然的亲近与舒适。仿古砖和汉白玉的运用打造出小桥流水的意境，配合传统壁纸及实木构造，让浓浓的中式韵味自然流露，无需创造。

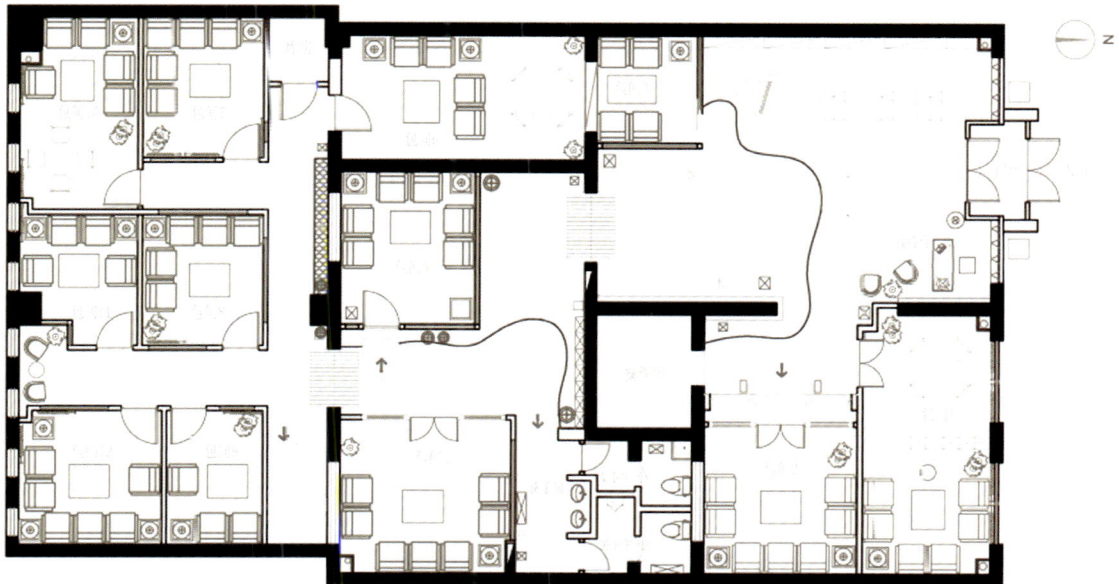

SHANGHAI JIANGNAN HUAFU PRESIDENTIAL MANSION CLUBHOUSE

上海江南华府总裁官邸会所

—— 地点
上海

—— 面积
1919 平方米

—— 设计师
周文胜

—— 设计公司
源创品格（香港）设计有限公司

—— 主要材料
酸枝木饰面、黄木纹石材、路易斯白石材、玛雅
米黄石材、帝皇金石材、彩虹木纹石材、紫罗红
石材、阿富汉黑金花石材

A PRIVATE FEAST
Entertaining Club
DESIGN

私享盛宴·娱乐会所设计

The clubhouse is located in Zhujiajiao, an ancient water town of Shanghai, and with the reputation of "the back garden of Big Shanghai" as its impressive geographical location, it's situated by the bank of the beautiful Dadian Lake, which highlights the upper class pursuit of value and quality. As a high-end private mansion clubhouse enjoyed exclusively, it's equipped with more than 10 sets of intellectualized facilities whose functions are a combination of international group conferences, party, red wine, body building, billiards, banquets, film and TV, and spa, which is a classic example of top luxurious houses among the first-tier cities in China.

The house is defined as a style of classic Shanghai Old Foreign-style House, whose age-old culture serves an index and enlightenment, with its inspiration drawn from the Shanghai classic and old architecture represented by "The Bund 18". In addition, by collecting and absorbing traditional European elements, putting the Chinese and western cultural essence into the design in a full way and making every effort to exhibit every detail perfectly, not only does the design demonstrate the luxury and magnificence of the space, but also skillfully renders a noble, splendid, and luxurious house with good taste as well.

本案地处上海水乡古镇朱家角，坐落于美丽的大淀湖畔，拥有着被誉为"大上海后花园"的傲人的地理优势，着重体现上流阶层对价值品质的追求。作为高端私人官邸独享会所，配备智能化设施十余种，集国际集团忙会议、聚会、红酒、健身、臬球、宴会、影视、水疗等多种功能于一体，是中国一线城市顶尖豪宅的典范。

本案定义为经典上海老洋房风格，以上海老洋房的文化底蕴作索引，从以"外滩18号"为代表的上海经典老建筑中寻找灵感，并搜集和吸收欧洲传统元素，将中西文化的精髓充分地运用到设计中，力求完美地表达每一处细节，不仅突出体现了空间的豪华大气，更巧妙地将整个空间晕染为高贵、华丽与奢华并重的豪宅品位之所。

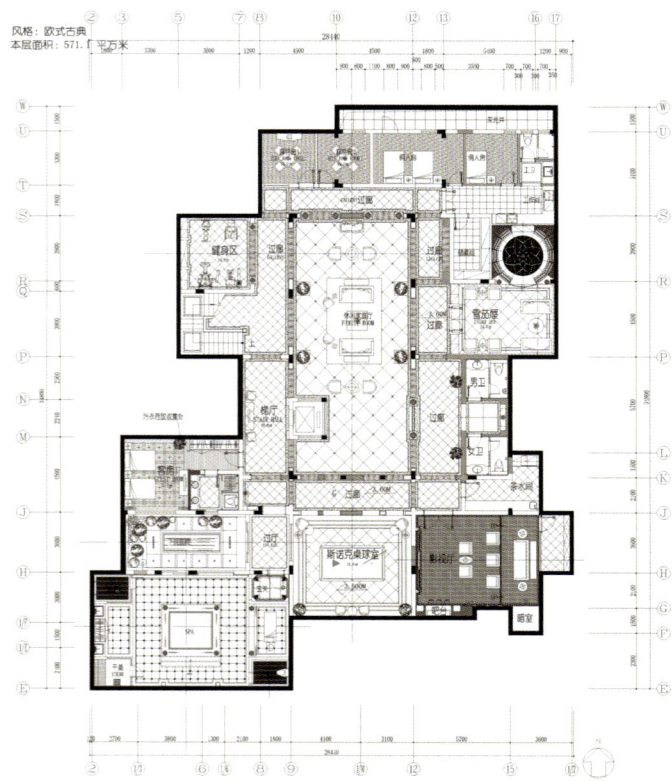

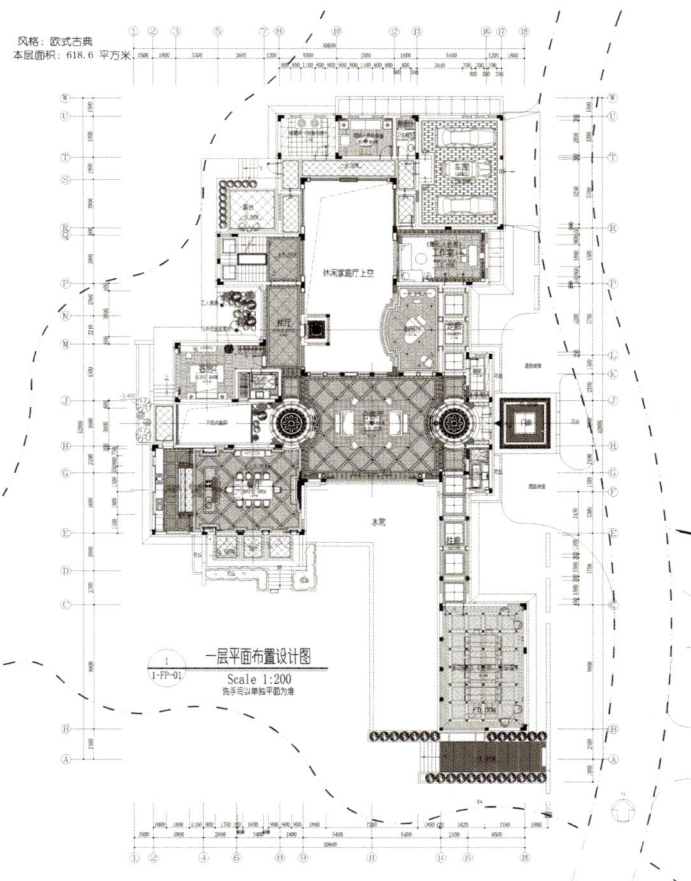

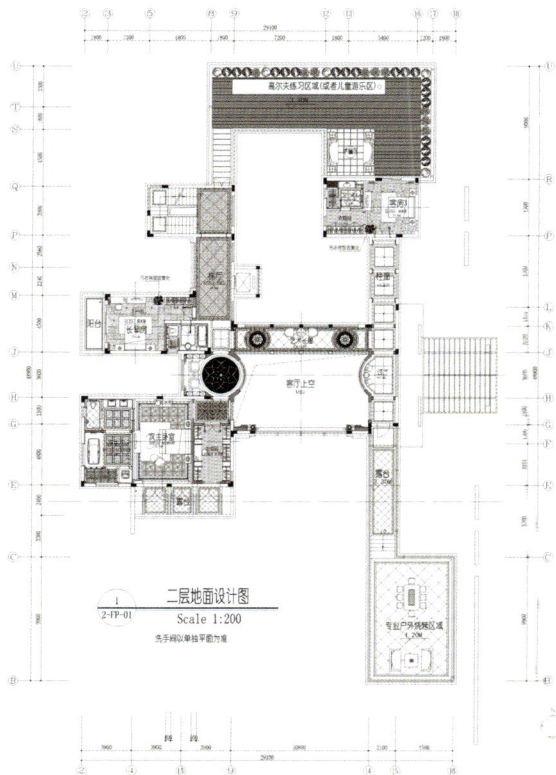

二层地面设计图
Scale 1:200

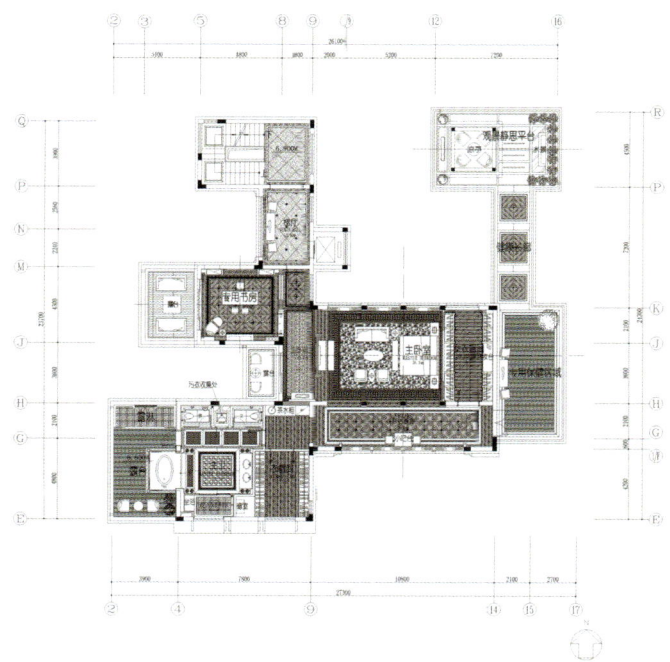

TIANYI MINGDUHUI CLUBHOUSE, ZHONGSHAN

中山天乙名都汇会所

— 地点
中山

— 面积
2000 平方米

— 设计师
周文胜

— 设计公司
源创品格（香港）设计有限公司

— 主要材料
新古堡灰石材、沙漠风暴石材、普罗旺斯石材、罗曼金石材、路易斯金石材、路易斯白石材、现代木纹石材、比利时棕石材、橡木、银箔

A PRIVATE FEAST
Entertaining Club
DESIGN

私享盛宴：娱乐会所设计

The shape of the main body of the whole architecture is an unsymmetrical "cross" with its horizontal part as a transmeridional architecture and vertically, starting from the gate, in the direction from north to south are the sale lobby, areas of car and piano, and as the oval-shaped hall extends north to the main model exhibition area and negotiation area, which are lined with film and video area, VIP area, bathrooms and offices, the axis of the space runs through the space of model exhibition area to the north until it reaches the end of the cross, where the water affinity platform is located. As the vertical part of the cross ends, the transition from an indoor enclosed space to outdoor open space is completed.

By locating the central axis, the design of the modeling re-divides and re-integrates the horizontal and vertical space to bring more sense of luxury and stereovision while highlighting the functionality of services, so that both the consumers and the staff can feel more about the humanistic beauty of the space on the basis of providing satisfying functional services.

整个建筑主体的形式是一个非对称的"十字"，十字的横向为东西向的建筑体，纵向从大门开始，自北向南开始通过销售大堂、酒吧区、钢琴区，并随着椭圆形过厅向北延伸到主模型区、洽谈区、东西两侧分别是影视区、VIP区和卫生间、办公区，空间中轴线通过模型展示空间向北一直到十字的终点——亲水平台，十字结束的同时，一个从室内封闭空间到室外开敞空间的过渡被完成。

造型设计上通过对空间中轴线的定位，将横向和竖向空间重新划分并整合，令空间本身的特质更具豪华感、层次感，并突出服务的功能性，在满足功能服务的基础上让消费者和工作人员更能感受到空间所具有的人文之美。

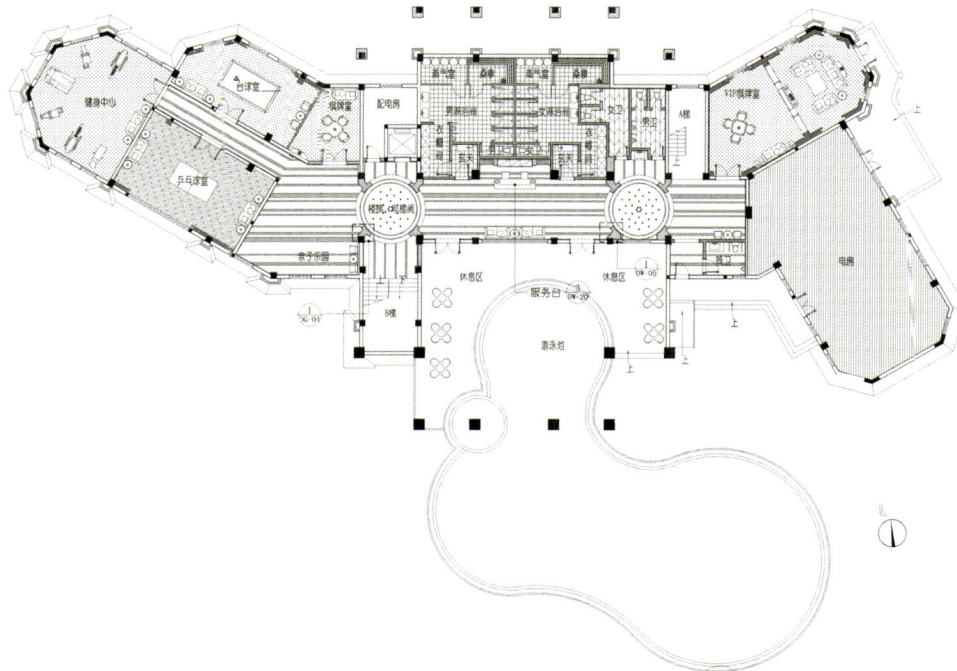

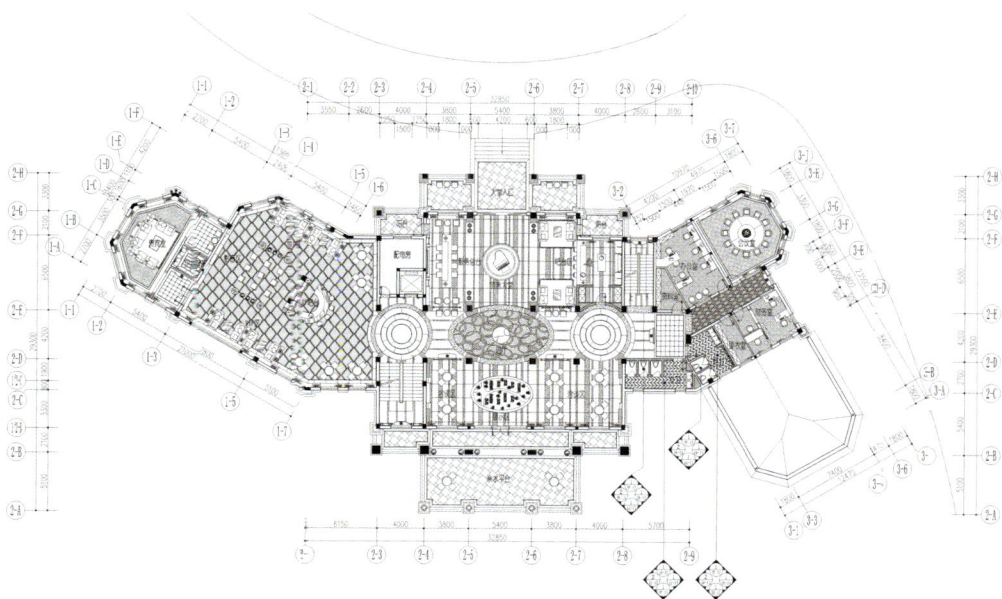

ORDOS GENGXIANG TEAHOUSE

鄂尔多斯更香茶楼

—— 地点
内蒙古

—— 面积
3000平方米

—— 设计师
李鹏

—— 设计公司
圣唐古驿室内设计事务所

A PRIVATE FEAST
Entertaining Club
DESIGN

私享盛宴：娱乐会所设计

Ordos, the new and emerging city located in the in and of China, is a city of coal, a city of wealth, but also a city lacking in culture. The investor builds Gengxiang Teahouse with the intension to inject elegance and freshness into this city. The site area of the project, more than 3,000 square meters, is divided into the tea-gift sale area, red wine area for exhibition and sale and catering compartments. Chinese and western styles are mixed in this design, with the western-style outline matching a Chinese-style texture. Mirrors are used to extend the space and stone is used for the wall and floor of the public area. The main color of furniture decorations is dark coffee, which appears to be elegant and solemn. Shuttling back and forth in the space, looking at the exquisitely built walls, doorposts and even the beam tops, touching the furniture reproductions with carved patterns of Ming and Qing Dynasties, all these bring people a deep nostalgic feeling, which leaves them immersed in the great cultural space to savor the fragrance of the tea and the mellowness of the years.

地处内陆的新兴城市鄂尔多斯是煤炭之城、财富之城，也是一座缺少文化的城市。投资方兴建更香茶楼，是想给这座城市注入一些风雅，一些清新。本项目建筑面积3000多平方米，分为茶礼品售卖区、红酒区展售和餐饮包间区。设计采用中西混搭的方式，西式的轮廓中搭配着中式的肌理。空间中多采用镜面拉伸空间，公共空间的地面及墙面材料多用石材，家具配饰以深咖啡色为主色调，典雅而庄重。穿梭在空间里，看着精雕细作的墙、门柱，甚至是梁顶，触摸着仿清明时期的雕花家具，带给人深深的怀旧感，让人沉浸在文化意味深厚的空间里，品尝茶叶的香浓，体味岁月的浓香。

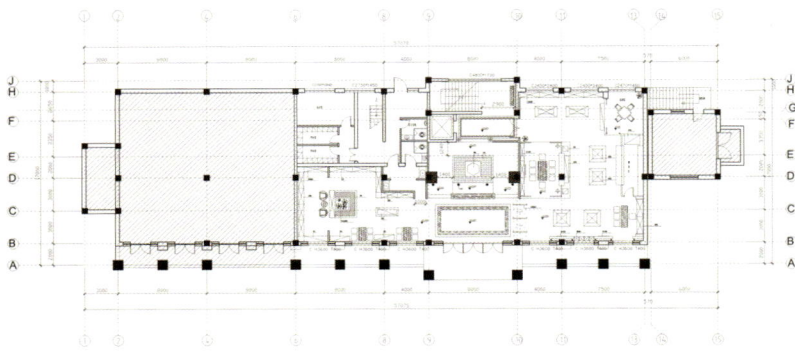

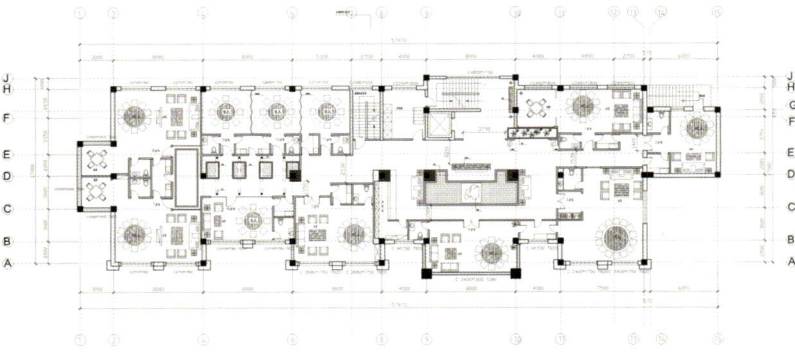

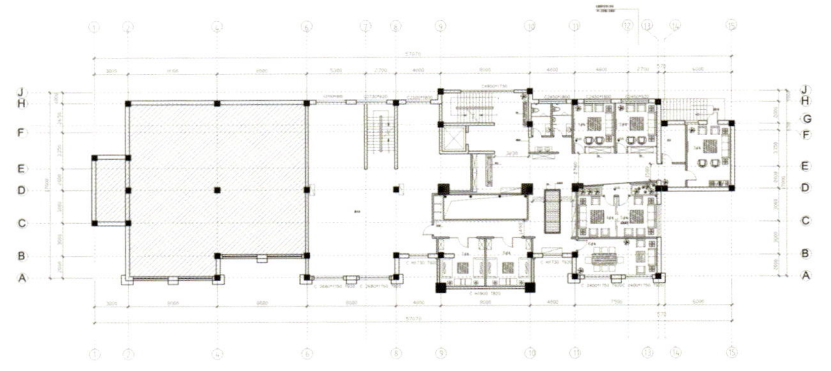

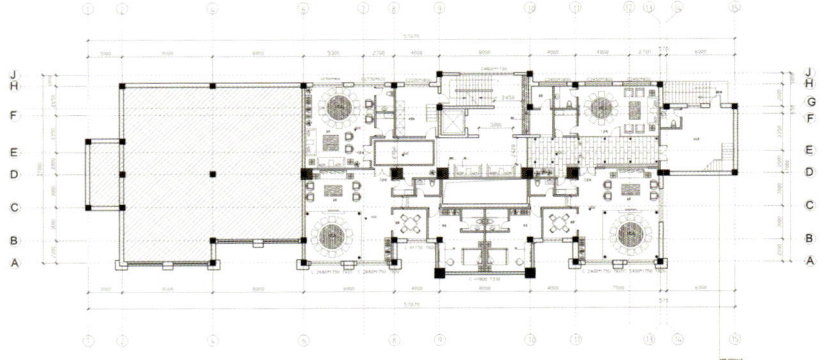

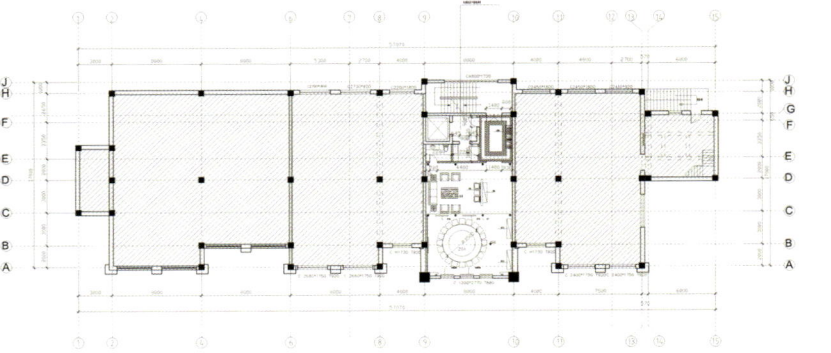

SICHUAN DUJIANGYANJADE SPLENDOR SENIOR BUSINESS CLUBHOUSE

四川都江堰玉垒锦绣高级商务会所

— 地点

成都

— 面积

3000平方米

— 设计师

黄河

— 参与设计

黄燚

— 主要材料

石材、不锈钢、铜饰、木饰面、地毯、墙纸、皮革、玻璃、乳胶漆、砖

A PRIVATE FEAST

Entertaining Club

DESIGN

私享盛宴：娱乐会所设计

Jade Splendor, a high-end business club in Dujiangyan, is located in Lidui Park, a National 5A-level Scenic Area. The club consists of two detached buildings, the first of which is mainly the high-class club, with the reception hall, the booths, the high-class business compartment rooms, and the grill rooms along the river, etc.. The second building is mainly the boutique hotel, containing sixteen high-class elite suites of various styles but each with a unique theme. The whole atmosphere as well as all the functions of the hotel carries a theme-culture connotation, in which the water-theme culture is melted into elements of the food, accommodation, trips, shopping and recreation offered by the club. Customers can both see the water and experience the culture of water. Meanwhile, everything in the club, including the environmental layout, building modeling, interior decoration, painting ornaments, furniture, lamps, staff uniforms, guest room supplies and recreational activities, conveys a connotation of water-theme culture, which makes the integrated design a complete, systematic, and meticulous one with both software and hardware.

都江堰高端商务会所玉垒锦绣坐落于国家5A景区离堆公园，共有两栋独体建筑。第一栋主要是高级会馆，设有接待大厅、卡座区、高级商务包房及沿河扒房等；第二栋主要是精品酒店，设置了16个高级、精品套房，每个房间风格不同，但都有一个特别的主题。设计将主题文化内涵有机地贯穿到整个酒店氛围和功能之中。把水主题文化融入到商务会所的吃住游购娱等要素中。让顾客既能看到水又能感受到水文化。同时在商务会所的环境布局 房屋造型、室内装修、书画装饰、家具 灯具、员工服装、客房用品、娱乐项目等各方面共同体现水主题文化的内涵，进行全面、系统、细致的软、硬件综合设计。

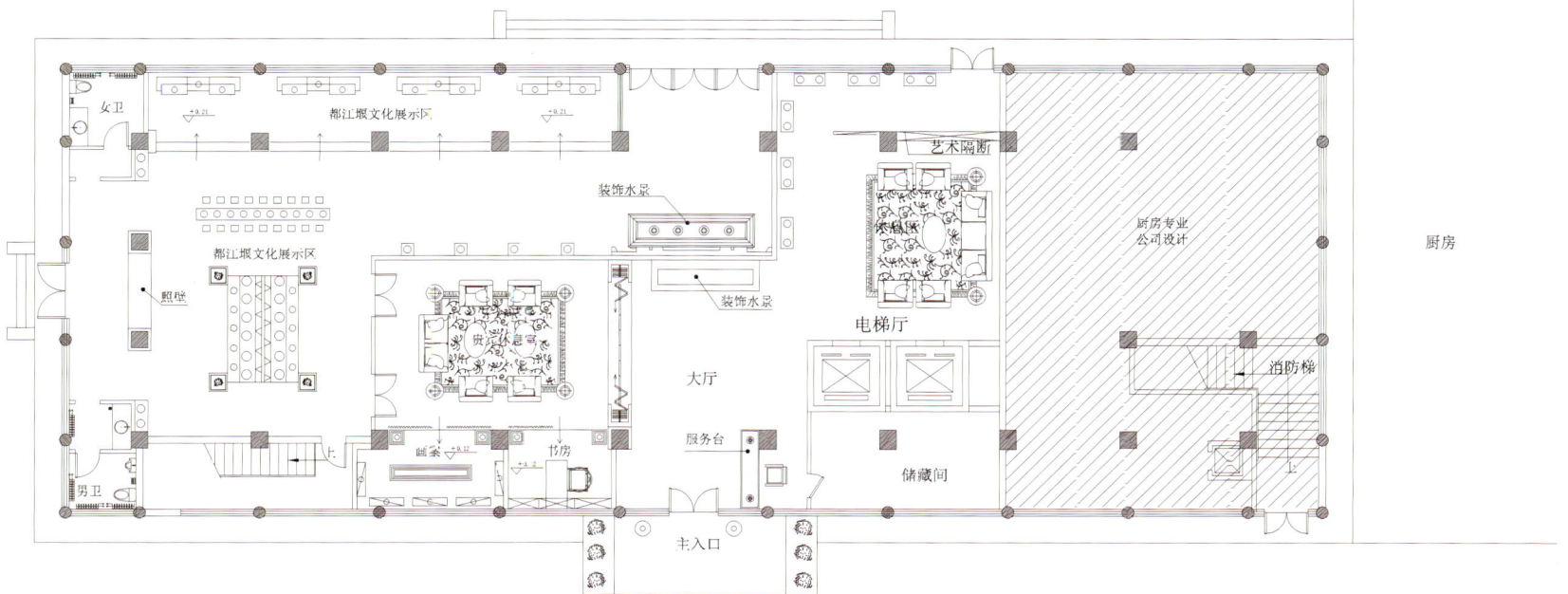

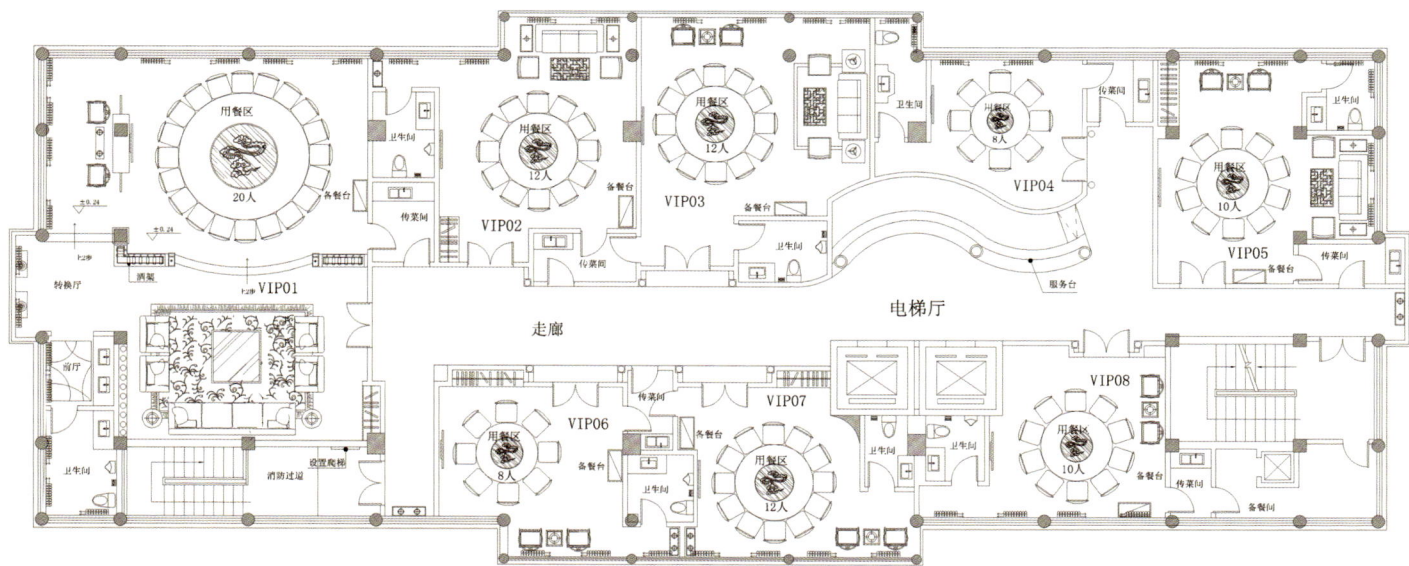

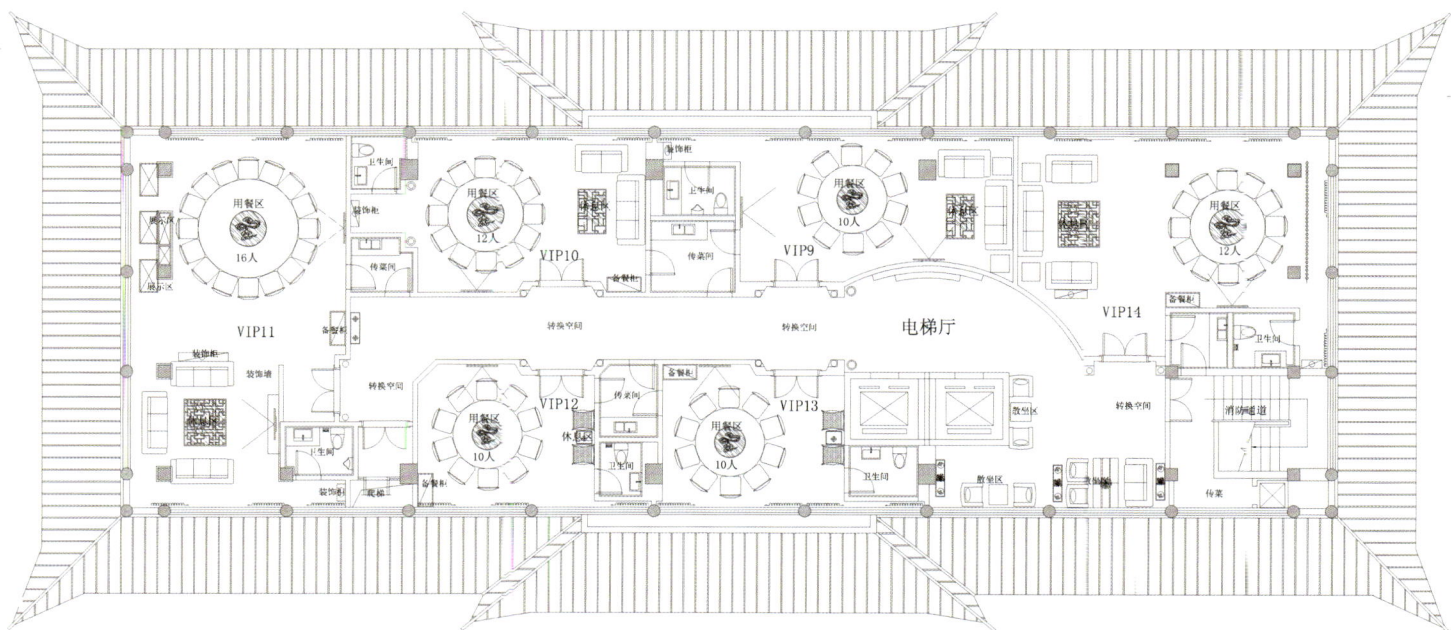

NIGHT CLUB 夜店

K TOP KTV

凯高点 KTV

— 面积
4000平方米
— 设计师
袁静、墨里黑
— 设计公司
朗昇空间设计
— 主要材料
大理石、地砖、玻璃等

A PRIVATE FEAST
Entertaining Club
DESIGN
私享盛宴：娱乐会所设计

Neo-classical style is mainly used in this project, coupled with the cool black and white, the bright red and blue and the different typical space modeling to create a recreational atmosphere with the theme of dynamics, fashion and prosperity.

Constructed around the original construction pillars, the CROWN booth in the lobby on the second floor becomes the visual focus of the entire lobby, splendid and powerful. CROWN is a function area including the characteristic drinking bar, the waiting area, the cashier's and the like.

The design on the third floor is a modeling of sky pillar, similar to the second floor but more honorable and luxurious in the space handling. Round the bar area, there are also the red wine room and the cigar room.

The president rooms and the big rooms make the best of the big scene to create atmosphere; and the small rooms are designed to be the variable rooms. Each of the rooms and the bathrooms within make use of different matches of colors to illustrate and enrich the modeling, which is economical and practical.

本案以新古典主义风格为主，运用冷峭的黑白色、艳丽的红蓝色以及不同的空间典型造型，表现出动感、时尚、繁华的主题娱乐氛围。

二楼大堂利用原有结构柱体为中心构造而成的"皇冠"卡座成为整个大堂空间的视觉中心、熠熠生辉，震撼人心。"皇冠"是充满个性特色的酒水吧、等待区、收银台等功能区。

三楼大堂则构造出一个擎天柱的造型，与二楼相仿，空间处理上要更尊贵奢华些，围绕酒吧区、还另设红酒坊与雪茄室。

总统房与大房间则是充分利用大场面来营造气氛；小房间则塑造成变化多样的包房。每个房间及内含的洗手间均利用不同搭配的颜色来表现，造型丰富，经济实用。

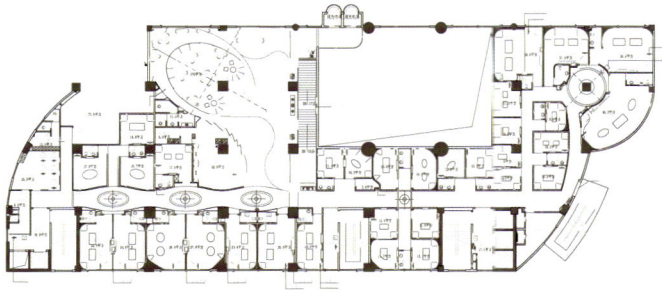

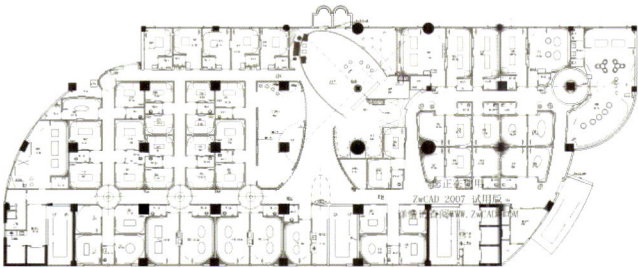

CONGA ROOM

康加俱乐部

— 地点
洛杉矶
— 面积
1300平方米
— 设计师
Hagy Belzberg

A PRIVATE FEAST
Entertaining Club
DESIGN
私享盛宴：娱乐会所设计

The Conga Room, in its new location at LA LIVE in downtown Los Angeles across from the Staples Center, is the city's premier Latin nightclub. The space will feature today's hottest Latin performers in its 1,300 square meters live venue space, which includes a restaurant, three distinct bars, patio seating, and a VIP lounge and private room. In addition, the club will host LA TV and world-renowned DJs adjacent to the stage and above the crowd, adding even more excitement to the ambiance. Perhaps the club design's greatest intent is to be true to the energy of the Latin community, to pay homage to its roots and deep history while infusing it with Los Angeles' fervent modern lifestyle.

康加俱乐部是洛杉矶市顶级拉丁夜总会，其新址位于斯台普斯中心对面，洛杉矶市中心的"洛杉矶活力"所在地。室内空间包括一个餐厅、三个特色酒吧、露天座椅区、一个贵宾休息室和一个私人房间，当今炙手可热的拉美演员将在其1300平方米的场地现场表演。此外，俱乐部会邀请洛杉矶电视音乐节目主持人和世界知名流行音乐主持人到舞台旁跟观众互动，给整个氛围增添更多兴奋感。或许俱乐部设计最大的意图是真实展现拉美人的活力，在将拉美风情融入到洛杉矶热情又现代的生活方式的同时，向他们的源头以及深厚的历史致敬。

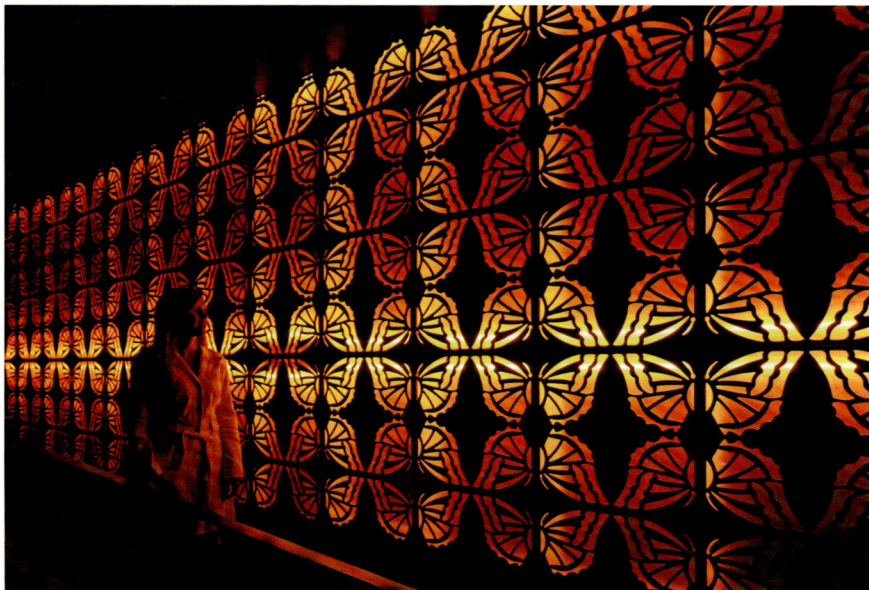

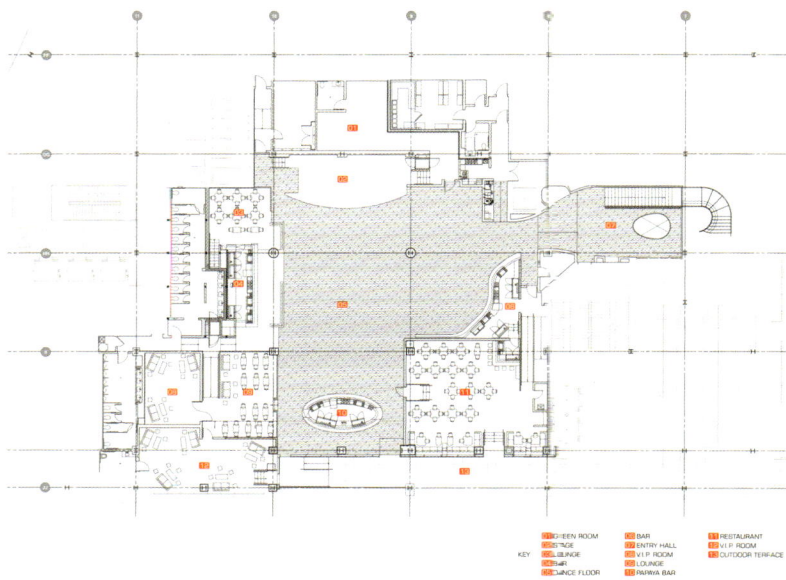

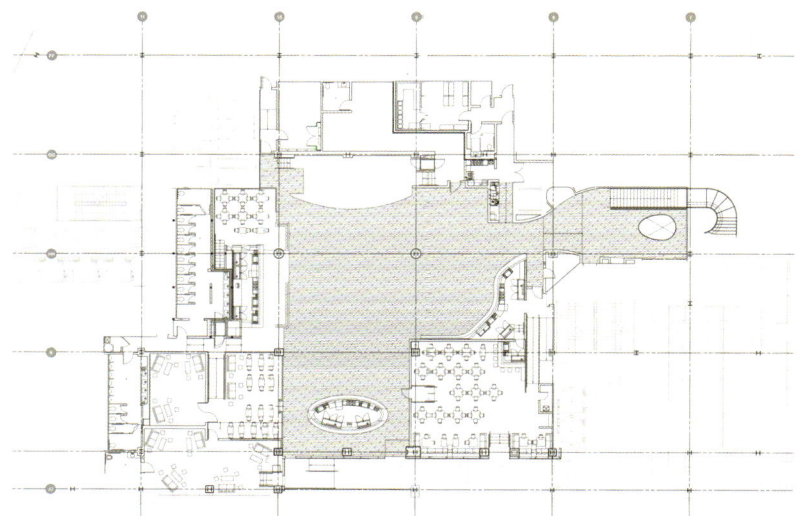

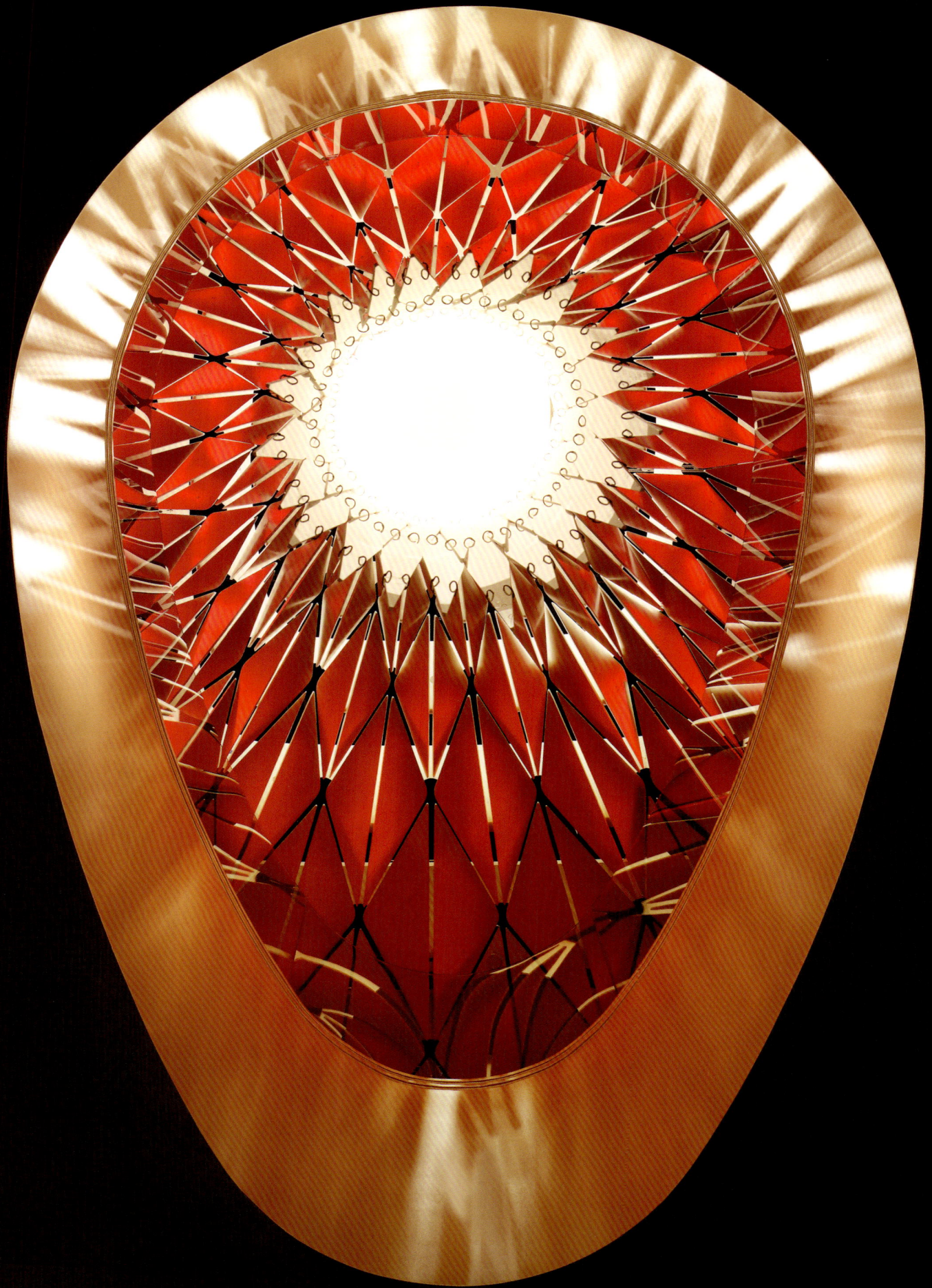

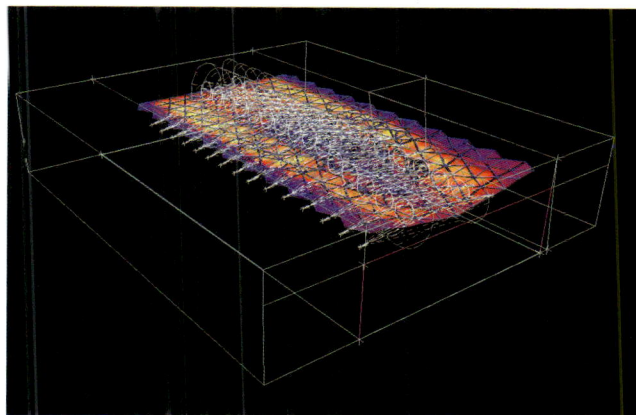
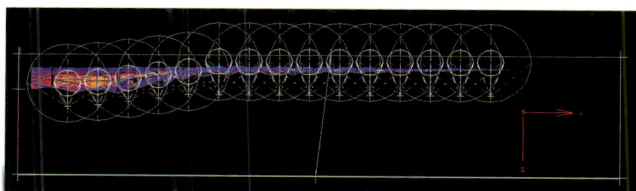

SHANGHAI LA LE WINE BAR

上海拉雷红酒吧

— 地点
上海

— 面积
170平方米

— 设计师
林琼然

— 设计公司
阔合国际有限公司

— 主要材料
黑色石材、黑镜、茶玻璃、镀钛不锈钢、柚木、
白橡木

— 摄影师
申强

— 业主
拉雷集团

A PRIVATE FEAST
Entertaining Club
DESIGN
私享盛宴；娱乐会所设计

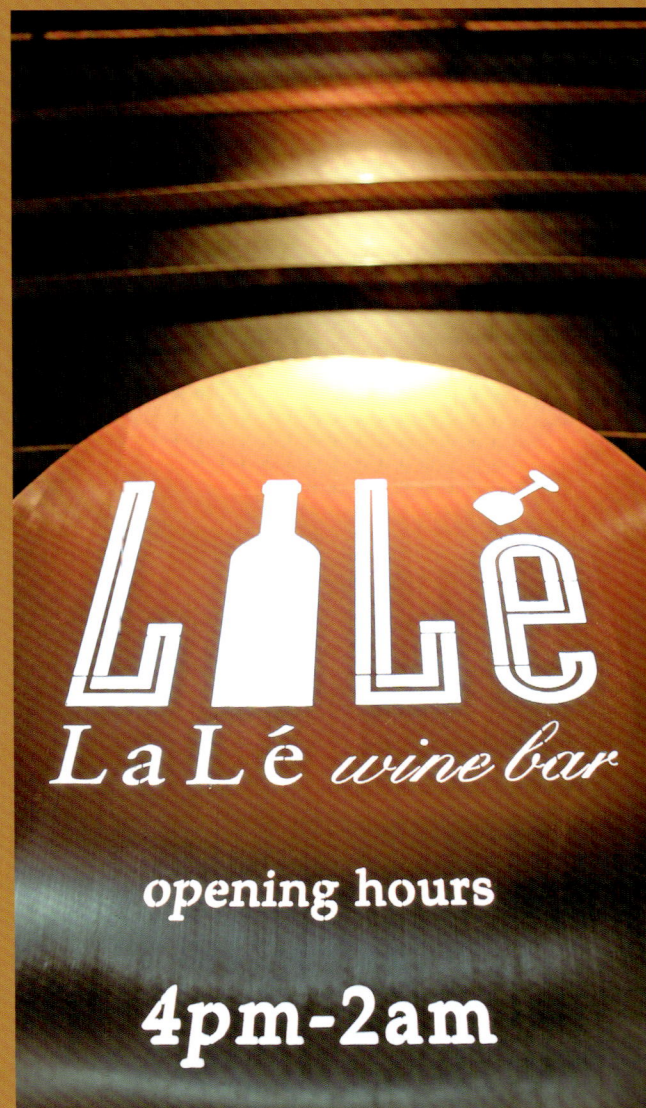

LaLé
LaLé *wine bar*

opening hours

4pm-2am

La Lé Wine Bar is a place where busy-urban people share casual evenings with friends at comfortab e ambience enhance with fine wines.

Evening Charm

The bar's black facade with fine metal curves making it a charm in the neighborhood. The sunset or moon like store sign visually loosen up tight shoulders. The relaxing vibes slows down pecestrians' footstep.

Gathering and Sharing

A bar is a field for interpersonal communication. The design interprets time and space providing city new comers a connection of urban experience and local communities that will further generage a new drinking culture. Customized furniture and light selection are arranged to suit demands in different areas separating bar area, wine tasting area and private rooms.

Modern and Surreal

The bar's aesthetic is a blend of modern and surrealism, articulated through the use of hard-edged materials like marble and glass meets local super class expectation. The curvy ceiling contour, wooden wall melody pattern and polished stainless steel creates the dizzy visual for a convivial vinous space.

Customized wine cellar with constant temperature and humidity control enables large collection of wines remains in good quality. Large glass viewing window makes it the center piece in the bar.

拉霍红酒吧是繁忙的都市人和朋友在舒适氛围中、由上等葡萄酒做伴，共享惬意夜生活之所。

夜幕中的魅力

酒吧黑色立面饰有精致的金属曲线，彰显其魅力所在。日落或如月亮般约店家招牌在视觉上缓解了人们的压力。轻松的氛围让行人放慢了匆匆的脚步。

聚会与分享

酒吧是一个人际交流的场所。这样的设计对时空进行着诠释，城市新人身处其中，可以体验到一种都市生活经历和地方社会生活的交融，进而产生一种新的品酒文化理念。定制的家具与灯具选择同时满足了不同区域人们的品位要求，酒吧也相应地被分为吧台区、品酒区和独立包厢。

现代与超现实

酒吧的美感在于通过像理石和玻璃这样的硬边材料的使用，实现了现代元素与超现实主义元素的融合，进而满足了人们对高品位的期望。曲线形天顶外观，木质墙体如音符般美妙旋律的图案和抛光不锈钢的运用为这个愉快的饮酒空间营造出一份视觉上约虚幻感。

量身定做的酒窖配有恒定的温度与湿度，使得在此库存约大量葡萄酒保持优良品质。大型玻璃可视窗的设置使其成为酒吧的中心区域。

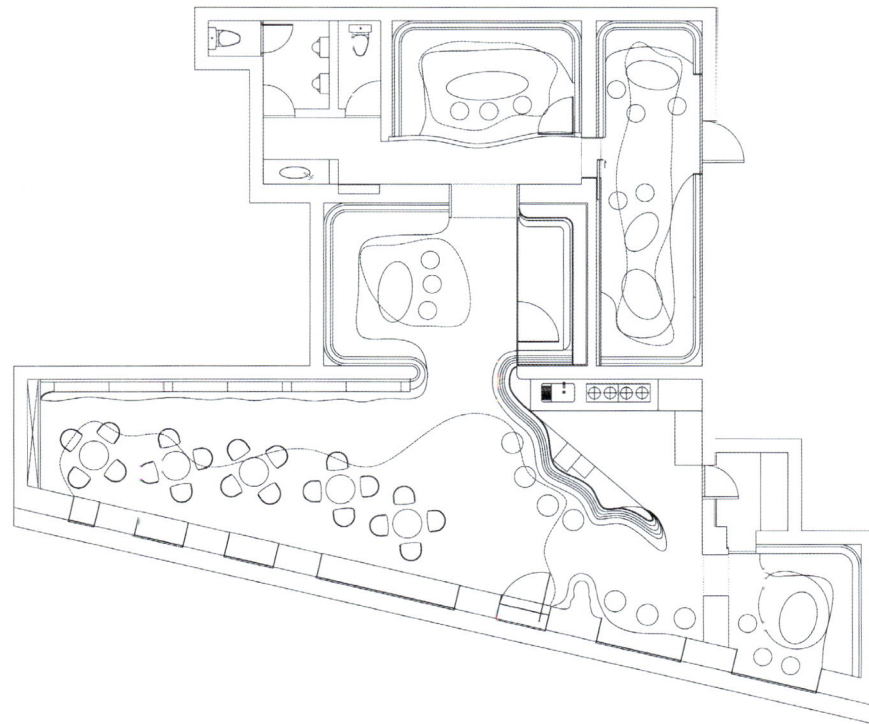

MELODY KTV IN CHAOWAI

麦乐迪KTV朝外店

—— 地点
北京
—— 面积
550平方米
—— 设计师
王俊钦、彭晴
—— 参与设计
赵文静
—— 设计公司
睿智汇设计
—— 主要材料
氟碳喷涂铝型材、不锈钢、LED、玫瑰金镜面
不锈钢、茶镜、麂绒布
—— 摄影
孙翔宇

A PRIVATE FEAST
Entertaining Club
DESIGN

私享盛宴：娱乐会所设计

The designer names this project Flying Colors, in which the blue seawater, the sinuous coast line, the sea waves and the undersea oxygen bubbles all become the designing elements and the whirling sea waves are the key point elements which run through the whole theme of the project. Design techniques like detachment, connection and incision are used to lead various modeling designs, connect or separate spaces, and generate mutual infiltration between the current design and the original space to realize an excellent match. In connection with the market positioning which aims at metropolis men and women as its consumer group, Flying Colors' lighting design also acts well in annotating and contrasting this idea. The LED lighting outlines the changes of the entire space and the reflecting finishing materials such as mirror stainless steel and glass emit light from all around.

设计师将此次设计方案命名为"流光溢彩"。其中，"蔚蓝的海水"、"蜿蜒的海岸线"、"海洋浪花"、"海底氧泡"都构成了设计元素。回旋起伏的"海浪"是贯穿主题的关键。运用了独立、串联、切割等设计手法牵引各种设计形态，连接或区分各个空间，并使现有设计与原有空间达到相互渗透、珠联璧合的状态。对于消费群体以都会男女为主的市场定位，"流光溢彩"的照明设计也很好地对此加以全释和映衬，LED灯光的运用将整体空间的变化勾勒出来，再通过镜面不锈钢、玻璃等反光好的装饰材质将灯光作多元化的敛射。

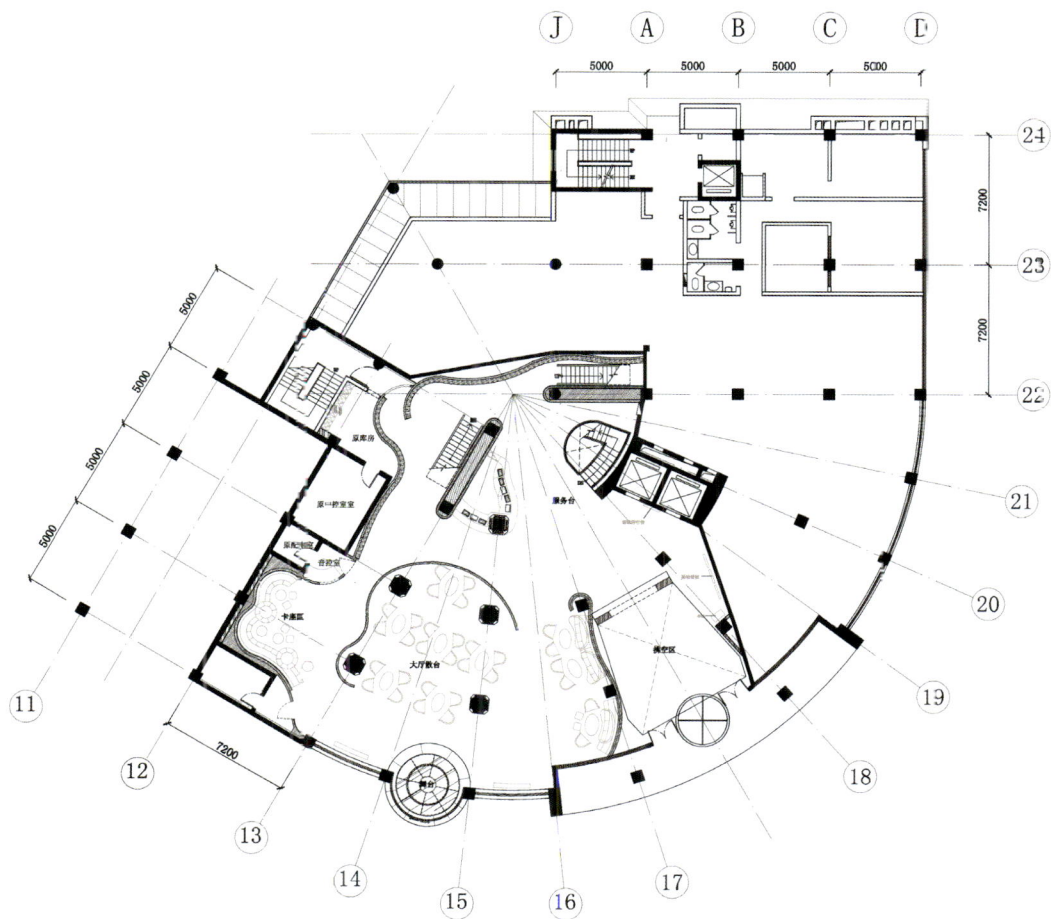

MELODY KTV IN ZHONGFU

麦乐迪KTV中服店

— 地点
北京

— 面积
3300平方米

— 设计师
王俊钦、彭晴

— 参与设计
赵文静、曹永辉

— 设计公司
睿智汇设计

— 主要材料
人造石材、石材、不锈钢、玻璃导板+LED、镜面不锈钢激光冲孔、激光雕刻喷涂、皮革、黑镜、防火板

— 摄影
孙翔宇

A PRIVATE FEAST
Entertaining Club
DESIGN

私享盛宴：娱乐会所设计

Melody in Zhongfu is located in Beijing CBD international trade business area, with a middle and high end market positioning and a consumer group of metropolis men and women. Based on this market position, the designers create a modern and fashionable place with embedded dream elements.

The stereo combination of the Meteor Shower suspended ceiling, the 3D LED metope, the Ice Brick and the metal-chain vision wall in the hall provides a dreamlike vision which fill people with admiration. The 3D LED lamplight vision wall in the middle of the hall is also a highlight. It is a show of universe played in the hall. The laminated combination of the LED lamplight and the acrylic together with the constantly changing shapes and colors, constructs a dynamic and magical astrospace. The metal chain modeling, contrasted with the suede nap as its backdrop, appears to be more distinguished and conveys another meaning of wealth sources. Desire and indifference, unconcern and dazzling charm, all these bring a beginning but never an end...

麦乐迪KTV中服店地处北京CBD国贸商业区，市场定位为中高端市场，消费群体集中于城市都会男女。设计师根据市场定位，创造出一个现代而时尚、具有内在梦幻元素的摩登场所。

大厅里的"流星雨"吊顶、3D LED 墙面、"冰砖"与金属环造型墙的立体结合，让人因梦幻之景而惊叹。大厅中间的 3D LED 灯光造型墙也是大厅中的亮点，它是大厅中间演绎的宇宙之景。用LED灯光与亚克力材料的层叠结合，不断变化的各种图形与色彩，营造出动感的奇幻宇宙空间。金属环造型在黑色鹿绒布的衬托下更显高贵，金属环阐述着财源之意，欲望与淡泊、漠然与惊艳，只有开始，却不见落幕……

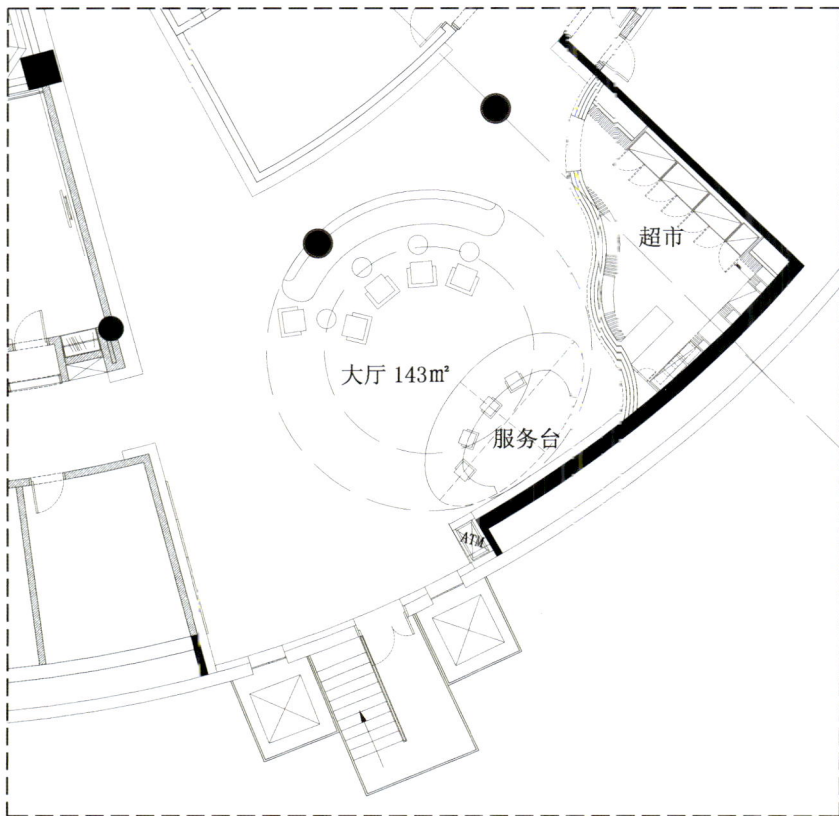

超市

大厅 143㎡

服务台

ATM

BARFOUFOU

醉醉吧

—— 地点
德国斯图加特
—— 面积
102平方米
—— 设计师
Peter Ippolito、Gunter Fleitz、Tim Lessmann、
Hakan Sakarya、Yuan Peng
—— 设计公司
OPPOLITO FLEITZ GROUP

A PRIVATE FEAST
Entertaining Club
DESIGN

私享盛宴: 娱乐会所设计

The FouFou is a champagne bar. You enter the FouFou straight into an L-shaped barroom arranged around a large bar counter. Mirrors affixed to the wall behind the bar expand the space further still. The bar itself is upholstered in a metallic greenish-gold diamond pattern, which is continued on the ceiling in the rear area of the barroom. The walls are executed in a complimentary shade of classic green. This forms an effective backdrop for the white, original case tree windows.

The Red Salon is located on the upper floor. The room is divided down the centre by two listed sections of wall that had to be preserved after refurbishment.

Bronzed mirrors mask this incision, creating a fascinating spatial perception. The ceiling is decorated by a lighting installation of semi-mirrored, golden light bulbs spanning the longitudinal axis of the salon, thereby conjoining both halves of the room.

Another salon with a refurbished wooden floor is located on the floor below. It is furnished with jet-black reproduction furniture commingling with modern elements in a metallic bronze tone. The individual elements are grouped to create intersecting ensembles of old and new.

醉醉吧是一家香槟酒吧。迈进酒吧，围绕巨型吧台设置的L形酒吧间尽收眼底。在酒吧后墙壁粘贴的大量镜面进一步扩充了空间范围。酒吧本身装饰的富有金属感的绿金菱形图案，一直延展到酒吧间后部的顶棚天花。墙壁施以淡雅的绿色，从而有效地衬托了新颖的白色框格窗。

红色沙龙厅位于酒吧上层。房间内一直到中心位置的部分由两段倾斜的、装修后必须保留的墙壁分隔开来。分割处覆以铜镜装饰，带来魅惑的空间感受。天花板上装饰的照明装置为半镜面金色灯泡，从而拉长了沙龙的纵轴线，让空间的两个半区结合起来。

另一个配有木质抛光地板的沙龙厅位于楼下位置。厅内陈设仿墨玉家具，混合金属铜色调的现代元素。单个元素经过精心组合，营造出古新交错的整体效果。

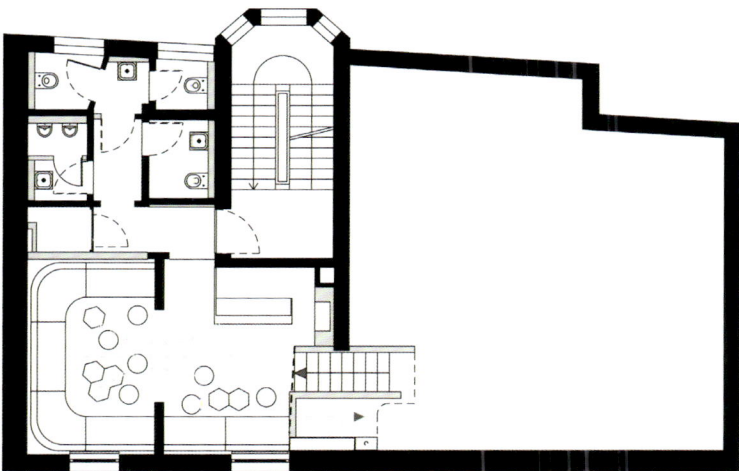

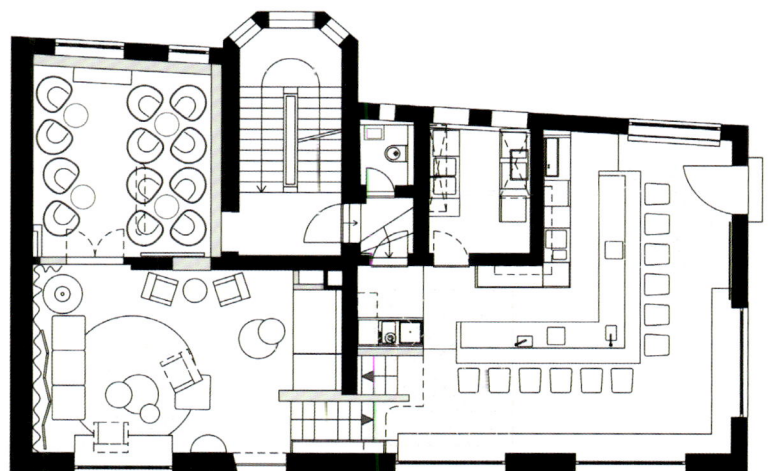

ROOM18 + 18LOVER

room 18 + 18 lover

—— 地点

中国台湾台北

—— 面积

850平方米

—— 设计师

甘泰来

—— 设计公司

齐物设计事业有限公司

—— 主要材料

压克力、黑色镜子、黑色EPOXY、木纹PVC、塑铝板

A PRIVATE FEAST
Entertaining Club
DESIGN

私享盛宴; 娱乐会所设计

Entirely based on the structure of letter L, the interior space of Room 18 is divided into two parts —18 Lover and Room 18, each with two interconnected entrances through which guests are free to walk.

By means of mock-up, 18 Lover attempts and considers the lines which must be laser cut on the acrylic classic board. Together with the LED lighting, it achieves the best effect of light, shadow and lines, and makes the classic elements "float" in the air in a brand-new way. It becomes the partly transparent partition between seats and thus a neo-British digital garden is shaped with illusive reflections and flowing lights and shadows.

Room 18 achieves the dramatic theatre effect by means of the high-low contrast. What surround the dance pool are the elevated compartment seats. The double-bar form is introduced, which shapes a multi-oriented and multi-leveled dance pool among the bars and the DJ booth and redefines the concept of a bar with the notions of "boy bar" and "girl bar". Here, the stage is everywhere you go and you are free to show the ebullient self.

整体L形架构的Room 18，内部共分为两个部分——18 Lover 与 Room 18，各拥有两处入口且能相互连通，客人可以随意游走。

18 Lover通过mock-up的施作，尝试与拿捏须镭射切割于压克力古典线板上的线条，再配以LED灯光的渲染，达到了最好的光影线条效果，让古典元素以全新的方式"飘浮"于空间中，使其成为坐席间半穿透的隔屏、进而塑造出虚实相间、光影变幻的新英式数位花园。

Room 18以高低阶差创造出了戏剧化的剧场效应，舞池四周是架高的包厢座位。透过一组双吧台的形式置入、使其和DJ台之间形成一多面向、多层次的舞池，同时，还衍生出新的吧台概念：男吧与女吧。在这里，随处是舞台、随时可展现奔放的自我。

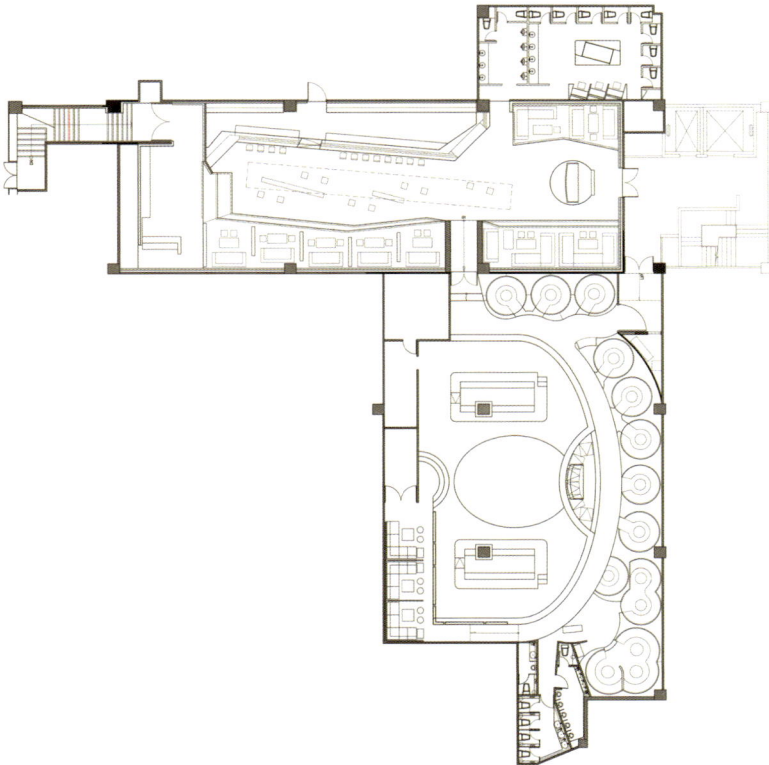

DROP SHANGHAI

上海Drop酒吧

— 地点
上海

— 面积
400平方米

— 设计师
Samuele Martelli　Giambattista Burdo

— 设计公司
欧比可建筑设计

A PRIVATE FEAST
Entertaining Club
DESIGN
私享盛宴：娱乐会所设计

DROP Shanghai: A contemporary classic.

With the definition of "contemporary classic" OOBIQ Architects refers to an interior made of classic textures, traditional shapes and materials, but transformed and adapted to a club to be located in such an international city as Shanghai.

That's why the interior started to have shapes that remind of European classic interior: boiserie—wooden panels with decorations—became the features walls all around the club, and classic leather sofa, so-called Chesterfield, became the outstanding seating of Drop.

For the bar and DJ booth: two sculptures of stainless steel in champagne finishing, where the three-dimensional decorations create vibrant and warm surfaces that will constantly reflect different tones of light.

The space is organized on different levels, with the VIP rooms and the seating areas located on platforms. The whole club is presenting the same floor finishing all around the areas: a geometrical Moroccan tile became the motif of the precious marble floor.

The warm decoration becomes another element that shows clearly the crossover of the interior design of Drop: elements of different cultures and times connected together in order to create a unique atmosphere.

上海Drop酒吧：一间当代经典酒吧。

欧比可建筑师对于"当代经典"的诠释是指，在室内设计中使用经典的肌理，传统的造型和材料，但对其加以改造以适合位于如上海这般国际都市的俱乐部的要求。

这就是室内设计为什么能够唤起对欧洲经典内部装饰回忆的原因：细木护壁板——配有装饰的镶板构成了整间酒吧的特色墙壁，冠名切斯特菲尔德的古典皮革沙发成为Drop酒吧里抢眼的坐席。

吧台和DJ卡座：两个香槟色的不锈钢雕塑配有立体感的装饰，营造出一种充满活力和温暖感觉的外观，不断反射着不同的光线色调。

空间的布局具有层次感，贵宾室和坐席区位于平台上。整个酒吧区铺设着统一的地面材质：一种几何摩洛哥瓷砖，形成了稀有大理石地面的主题。

温馨的装饰成为了表现Drop酒吧内部设计风格转变的另一元素：不同文化和时代的元素汇聚一体，营造出的是一个独特的氛围。

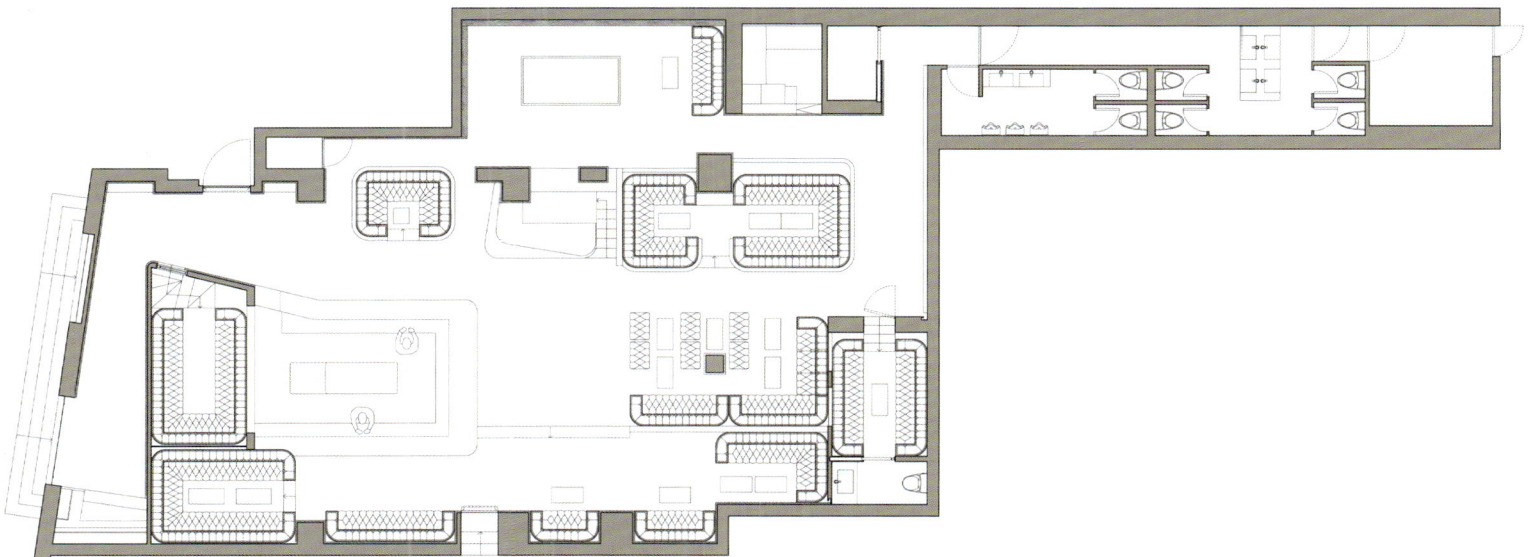

TIANZUN RECREATION CLUBHOUSE

天尊娱乐会所

— 地点
福建

— 面积
约2000平方米

— 设计师
金舒扬

— 方案审定
叶斌

— 设计公司
福建国广一叶建筑装饰设计工程有限公司

— 主要材料
镜面不锈钢、镜面、云石灯片、大理石、软包等

A PRIVATE FEAST
Entertaining Club
DESIGN

私享盛宴: 娱乐会所设计

Drum music and light dance, like in a fairyland; drinking wine by music, lingering in it. The evening-show place as a space where people search for unconventionality should provide people with unique experiences: serenity and desire, pressure and relief, weightless culture, lighting hints... and all these send enthusiastic calls to people. Getting absorbed, intoxicated, or indulged are all the feelings in one's own heart.

At the entrance, magnanimous and glaring step lights create a dynamic feeling of suspending and ascending, which brings an exclusive honorable enjoyment. Entering the interior space, the use of reflecting materials serves as a foil of the magnificent space to make it brighter and more absorbing, which is like a hint or temptation, but reminds people to remain elegant as well. This is a space with both elegance and captivation.

鼓乐曼舞如仙境，当歌对酒竟流连。夜场作为一个人们寻找常态之外的空间，就应该给人带去不同寻常的体验：平静和欲望、挤压和释放、失重的文化、光线的暗示……都向人们发出热情的召唤，或沉浸、或陶醉、或沦陷，都在自心。

入口处恢弘炫目的台阶灯营造出一种悬浮上升的动感，带给人一种专属的尊崇享受。进入室内空间，反光材料的运用将华丽大气的空间衬得更为明亮迷离，仿佛是一种暗示，又像是一种诱惑，却又提醒着人们保持优雅。这是一个集优雅与魅惑为一体的空间。

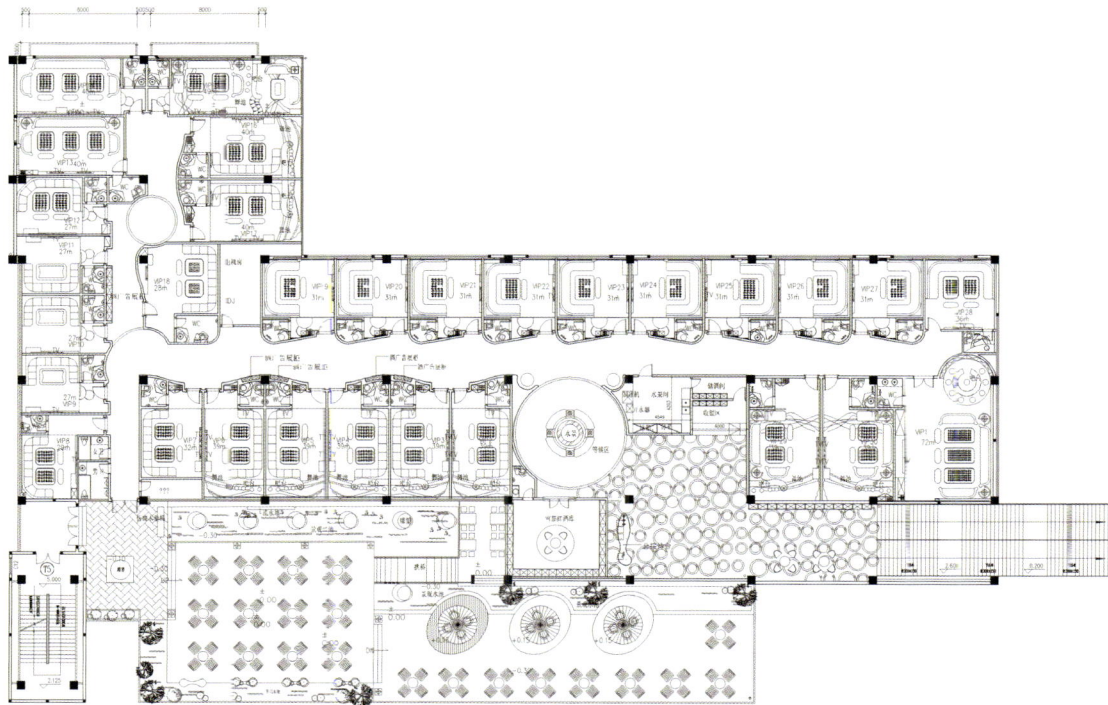

FLY ME TO POLARIS

星愿

—— 地点
南京
—— 面积
300平方米
—— 设计师
李浩澜
—— 设计公司
南京浩澜设计事务所
—— 主要材料
灰镜、马赛克

A PRIVATE FEAST
Entertaining Club
DESIGN
私享盛宴：娱乐会所设计

The Fly Me to Polaris Restaurant, only 300 square meters, lies silently at a corner of a flourishing pedestrian street without a glaring light box door or eye-catching entrance. The designer watched the movie, Fly Me to Polaris, when he was young and was moved to tears. Thus he wants to build this restaurant, the name of which is the same as the movie, into a romantic one to move people just as the movie used to. He even maps out scenes of marriage proposals, and without any blueprints at hand, the construction of the restaurant begins. Since its completion, the restaurant which has been operating for a long time is always welcoming lovers to spend the happy time here.

　　静静地躺在都市最繁华的步行街的一个角落，没有华丽炫目的灯箱门头，没有张扬醒目的入口，这就是仅有300多平方米的星愿餐厅。年轻时看《心语星愿》，感动得落泪。于是，想把这个同名餐厅也打造成一个浪漫的爱情餐厅，像《心语星愿》一样让人感动；甚至连一幕幕求婚场景都规划在其中，在没有图纸的情况下这个餐厅开工了。收工至今，已经营业了许久的星愿餐厅，无时无刻不在迎接着心中充满爱情的情侣在此共度美好时光。

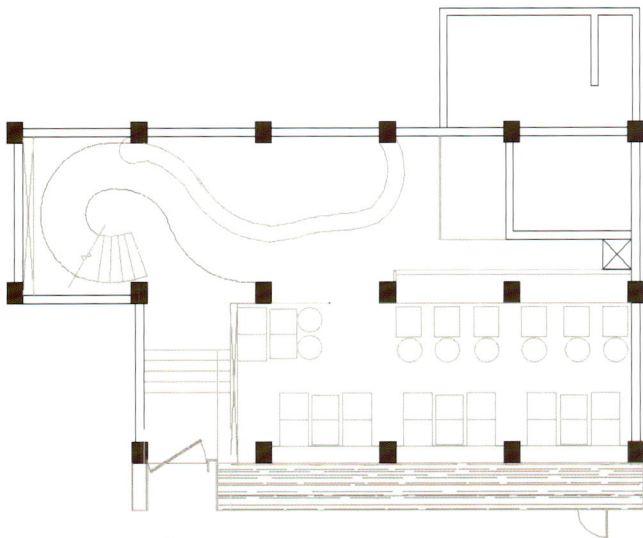

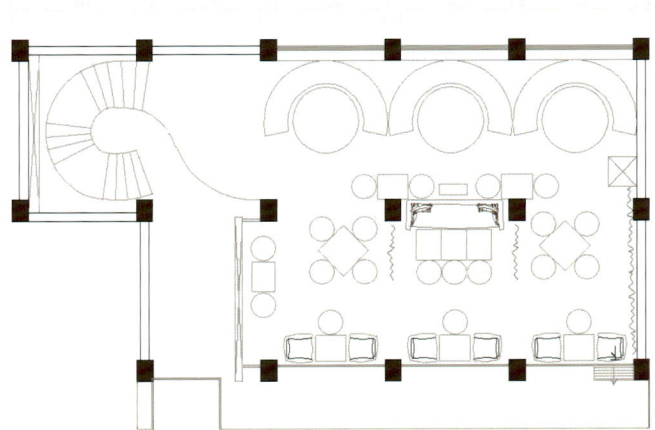

LAS VEGAS, SHANGDONG

山东拉斯维加斯

— 地点
山东
— 面积
4200平方米
— 设计师
林兰涛
— 设计公司
深圳阿凡达室内设计有限公司
— 主要材料
冷光玻璃钢、GRG、玻璃、定制砖、亚克力、水
晶、软包、LED、舞台特效灯光

A PRIVATE FEAST
Entertaining Club
DESIGN

私享盛宴: 娱乐会所设计

By naming the project "Las Vegas", the nature of the space is defined as a diversified leisure place. The designer takes local consumption level, geographical location, economic basis and the competition between rivals in the business into consideration and makes out the blank area as the design basis, adapts a diversified design concept, and gives clear distinction to the high, medium and low consumption area step by step.

The space is divided into five sections, named A, B, C, D and E respectively. It includes the FIT area, high-end reception area, party entertainment area, VIP member area and the terminal service area. Each area is designed into different style according to the different consumer groups and their demands, and each carries its unique characteristics while remaining unified. In lighting treatment, it takes the form of interaction, the lights move while people do so and when people stop, the lights stops, too. Meanwhile, it also uses light, electricity, cloud and fog as the media for lighting to shape the design and finally create a world of light and shadow in its true sense.

本案命名为"拉斯维加斯"的同时，也将空间的性质定义为多元化的消费场所。设计师结合当地的消费水平、地理位置、经济基础及同行业竞争对比、理出空白区域，作为设计依据，采取多元化的设计理念，逐步把高、中、低档消费区区分开来。

空间被分为ABCDE五个区域，包括散客区、高端接待区、派对娱乐区、VIP会员区和终端服务区。各个区域根据不同的消费人群和消费需求设计成不同的风格，让每个区域都有自己的特色，却又都统一在一起。灯光处理上采取互动形式，人走灯走，人停灯停；同时运用光、电、云、雾作为灯光的媒介来塑造设计，打造出了一个真正意义上的光影世界。

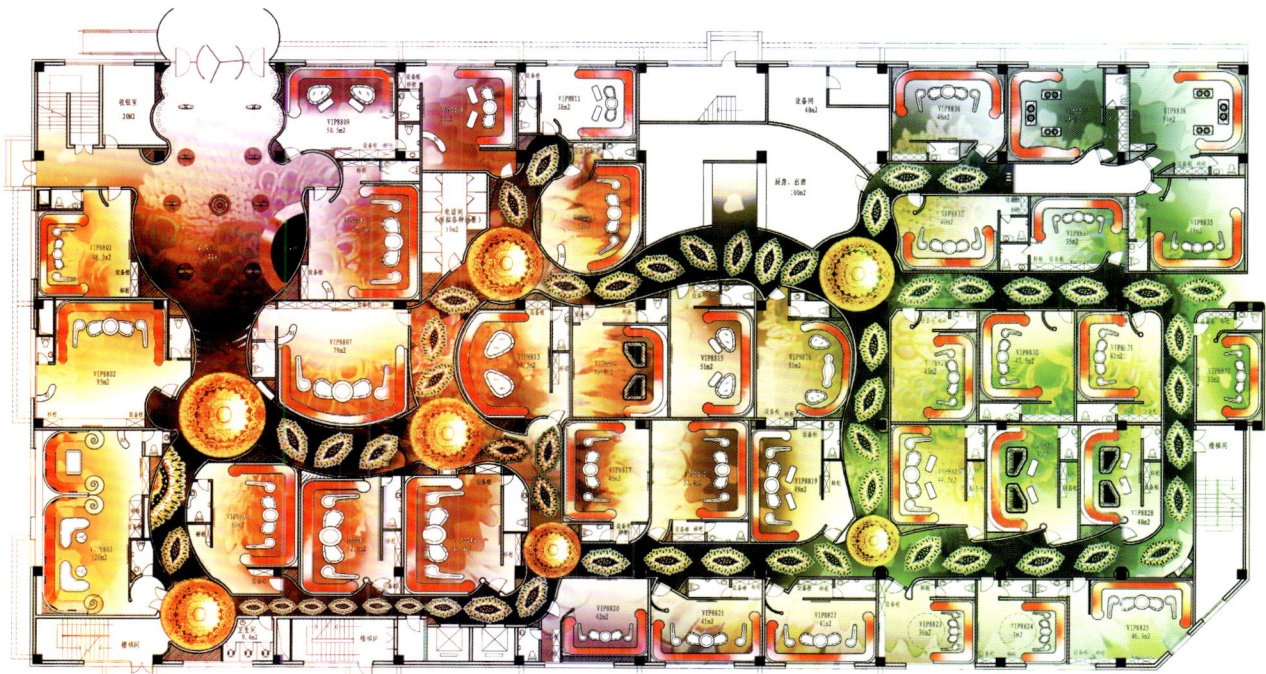

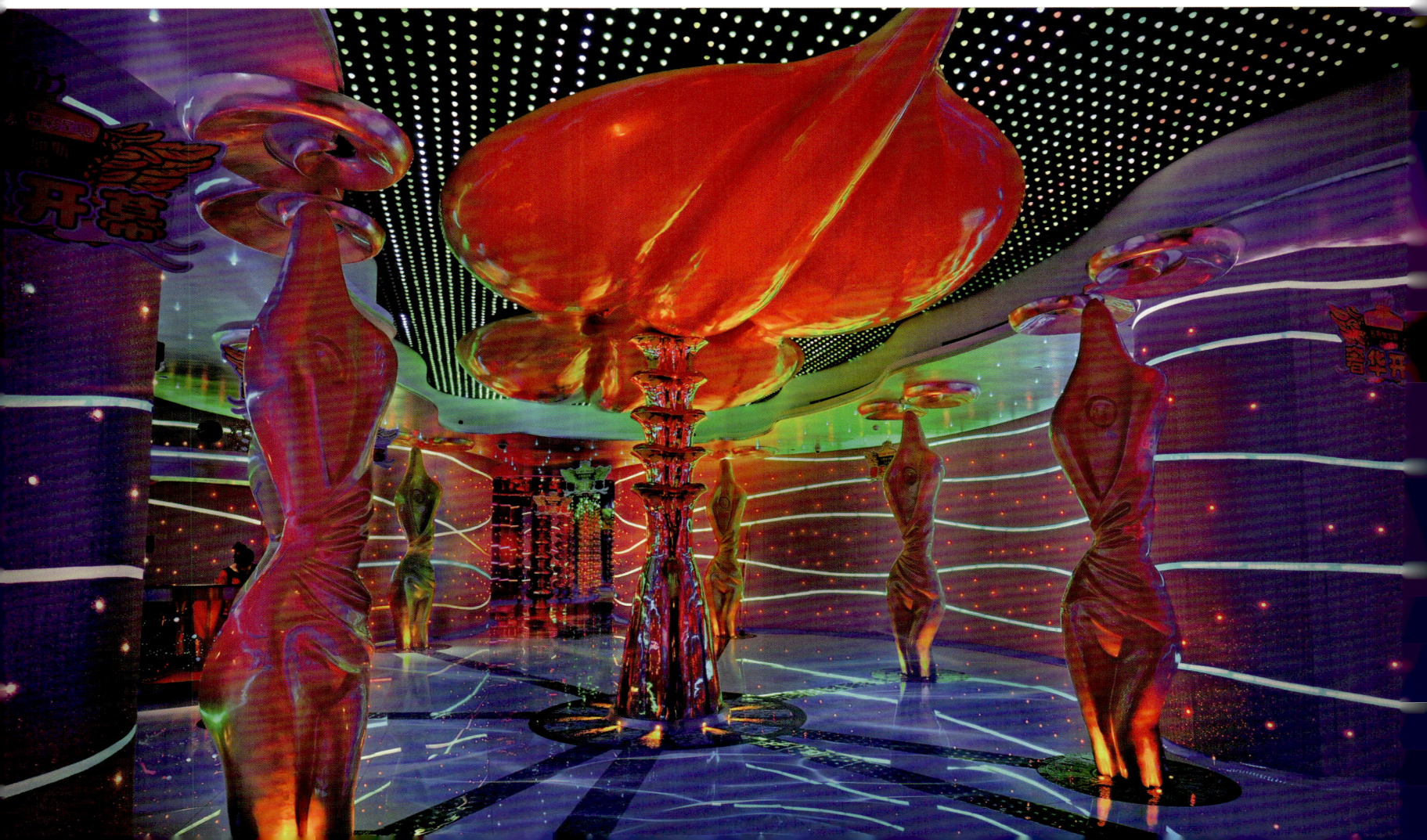

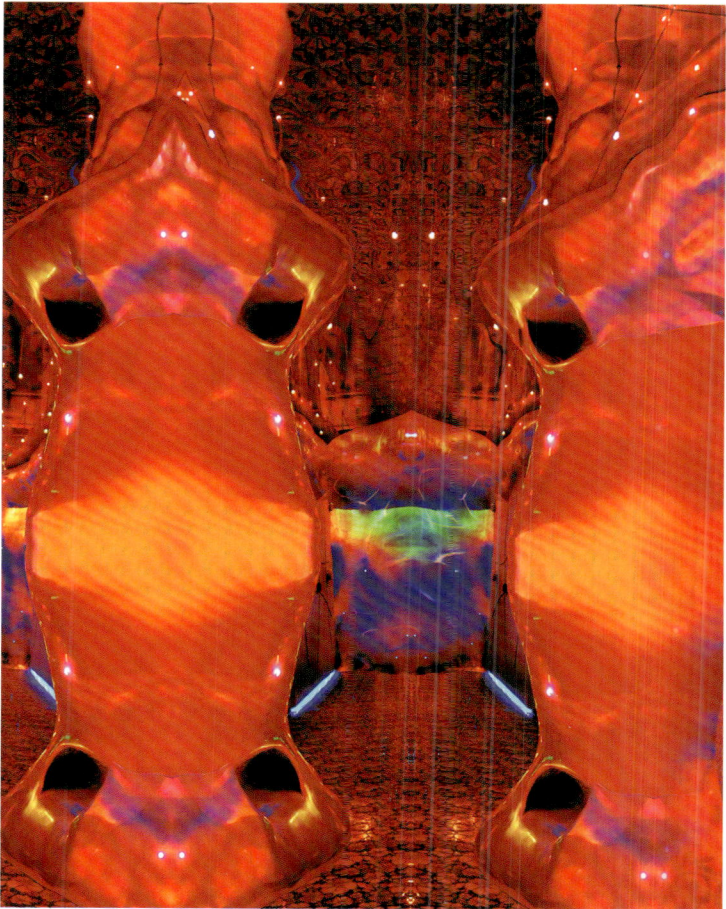

UNFETTERED IMAGINATION KINGDOM

畅想国度

— 地点
深圳

— 面积
4000平方米

— 设计师
吴恙

— 设计公司
深圳市吴恙室内装饰设计有限公司

— 主要材料
玻璃、金属

A PRIVATE FEAST
Entertaining Club
DESIGN

私享盛宴:娱乐会所设计

The project creates a colorful and entertaining space with modern sectional materials and LED lights. The designer pursues the combination of light and shadow to bring people endless imagination. In the design of aisles, materials with excellent reflective properties like marbles, PU, metal trimming and LED are integrated to create a blurring and tempting space which allows people to experience the sense of modern fashion.

Sound-absorbing upholstery and lighting design of the compartments are used to create a captivating atmosphere of dark colors, which adds a sense of mystery to the space. Moreover, the lighting techniques are the most important thing, because LED lamps play a vital role in this space. The changes of the lighting not only offer feelings of blurring and confusing, but also produce various sensory experiences for the space. The same space brings different experiences, and that is the keynote of "unfettered imagination".

本案运用现代型材和LED灯，为人们创造了一个五光十色的娱乐空间。设计师力图通过光与影的结合，带给人无尽的畅想。在走道的设计中，大理石、PU材料、金属镶边及LED等反光性能较好的材料综合在一起，创造出一个迷离、诱惑的空间，让人充分体验现代时尚感。

包房利用吸音软包和灯光设计营造出一份暗调的魅惑，增添了空间的神秘气息。此外，灯光的处理也是重中之重，在这个空间里，LED灯起着至关重要的作用，变幻的灯光既带给人迷离、错乱的感觉，同时也带给空间不同的感官体验，同一个空间带给人不同的感受，这也是"畅想"的主旨思想。

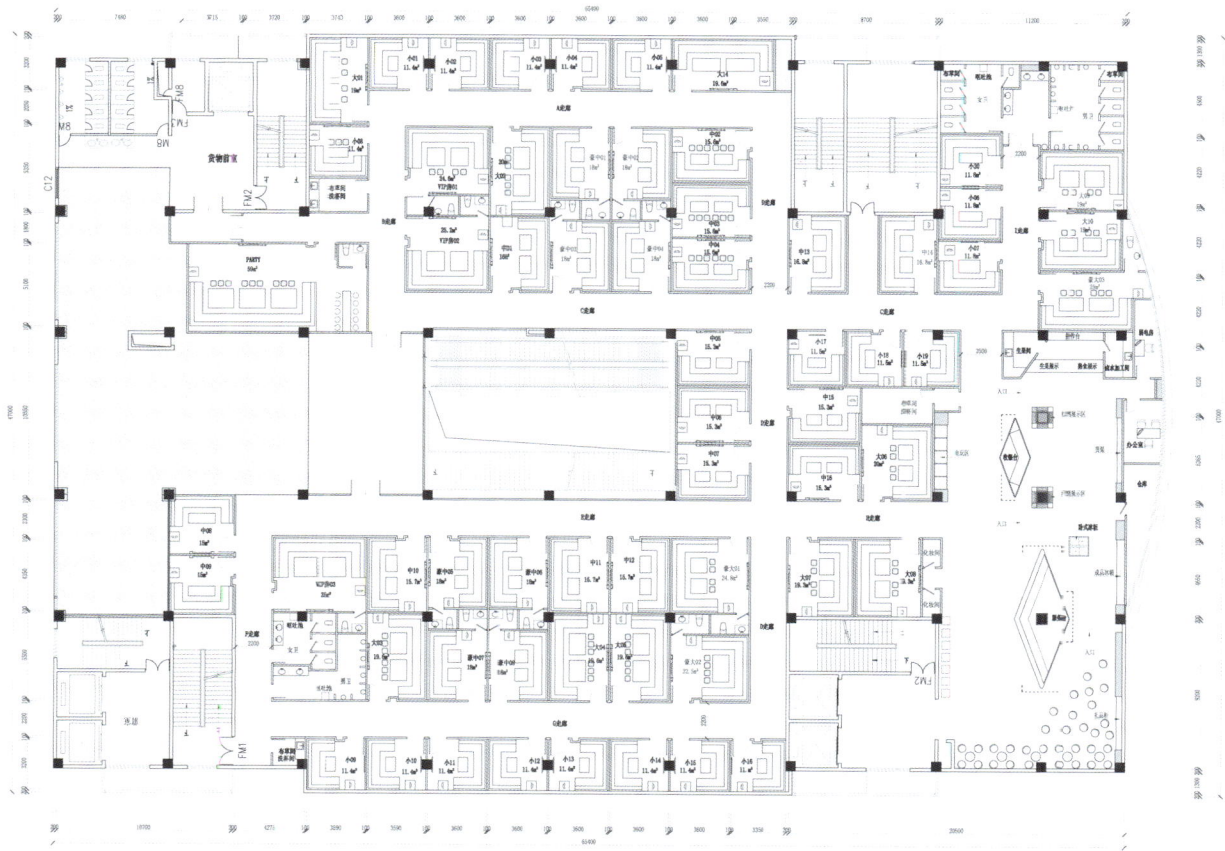

OUBANGHONG

欧邦红

—— 地点
深圳

—— 面积
500平方米

—— 设计师
吴恙

—— 设计公司
深圳市吴恙室内装饰设计有限公司

—— 主要材料
镜、防火板

A PRIVATE FEAST
Entertaining Club
DESIGN

私享盛宴·娱乐会所设计

Nowadays, with the social development and the increasingly fierce market competition, the luxurious style of extravagancy has been abandoned in the design of a leisure site. It is argely diverted to the pursuit of modern and fashionable features. In order to satisfy the ever-changing taste of the consumers, the designer therefore creates a place which is high-style and elegant.

At the entrance, large glass doors make the indoor scenery appear indistinctly.

Metal and glass match together to make the space display a cold temptation. In the design of the hanging staircase and the structured cellaret, straight lines are used to enrich the space form, meanwhile create a three-dimensonal effect and multiply spatial expressions. In lighting treatment, mild and dark tones are adopted to fill the space with elegancy and warm temperament. A little blurred, a little fantastic, and a little elegant, all make everything happen in a natural and sincere way.

随着社会的发展，在市场竞争愈发激烈的今日，休闲场所设计也开始摒弃纸醉金迷的奢华风格，更多地转为追求现代时尚风格，为了配合消费者不断变化的消费品味，设计师在此打造出了一个高格调的优雅场所。

在入口处，大幅玻璃门让室内风景若隐若现，金属材料与玻璃相搭配使空间呈现出清冷的诱惑。悬挂式楼梯、规整的酒柜设计用笔直的线条丰富着空间形态，同时营造出了立体感，增添了空间表情。灯光选择清淡、暗哑型的，既显优雅，又让空间充满温馨气质。一份迷离，一份梦幻，一份优雅，让一切发生得自然而真实。

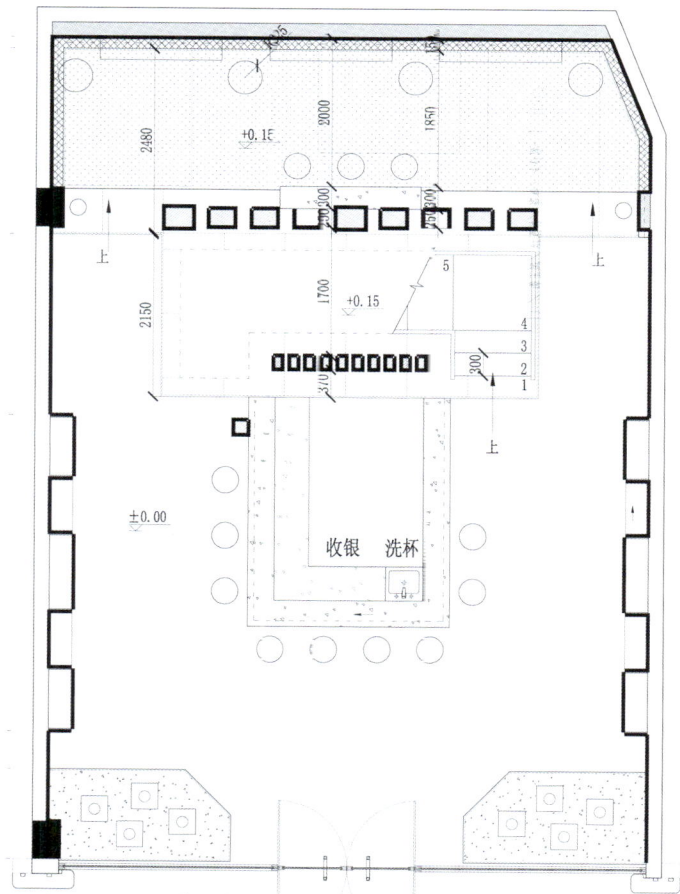

收银　洗杯

SOUND OF MUSIC
RETAIL KTV

音乐之声量贩KTV

— 地点
山东
— 面积
2500平方米
— 设计师
王超
— 设计公司
（卓尔）超设计事务所
— 主要材料
不锈钢、彩镜、光纤、大理石

A PRIVATE FEAST
Entertaining Club
DESIGN

私享盛宴：娱乐会所设计

In this project, light plays a major role. Bright light with subtle pattern conveys a warm welcome to the consumers. Whether immersed, intoxicated, or indulged, every feeling exists in the customer's heart. In order to echo the theme, the designer creates a magnificent atmosphere from the very beginning by the bold combination of colors, lights and the splendid form throughout the design. The wall surface and the ceiling are all cut into ever-changing geometric forms, and together with the shining pattern on the ground, all these make people to truly enjoy the musical rhythm and movement, as if it can call out another self hiding deeply in the heart and release passion and vitality even unfamiliar to youself. This is both the main theme of the designer and the glamour of "the Sound of Music" as well. In this space where technology melts into art, classic elements melt into modern ones and management ideology melts into design of the space, people can enjoy the transcendental noble temperament and the amiable expression, and are more likely to accept and participate in it.

本案中，光线是主角。炫彩的灯光加上诡秘的图案，都向人们发出热情的召唤，或沉浸、或陶醉、或沦陷，都在自心。为了呼应主题，设计师一开始就营造出一种恢弘的气势，大胆的色彩与灯光的组合以绚丽的形式贯穿整个空间。墙面、吊顶都以丰富多变的几何形式切割，与地面炫动的花纹搭配，让人真正感受到音乐的旋律与动感，仿佛能召唤出潜伏在心底的另一个自己，释放出让自己都陌生的激情与活力。这既是设计师的主旨思想，也是"音乐之声"的魅力，让人们在这个科技融入艺术、古典融入现代、经营理念融入设计的空间里，感受到空间超然的贵族气质和平易的表情，让人们更容易接受并参与其中。

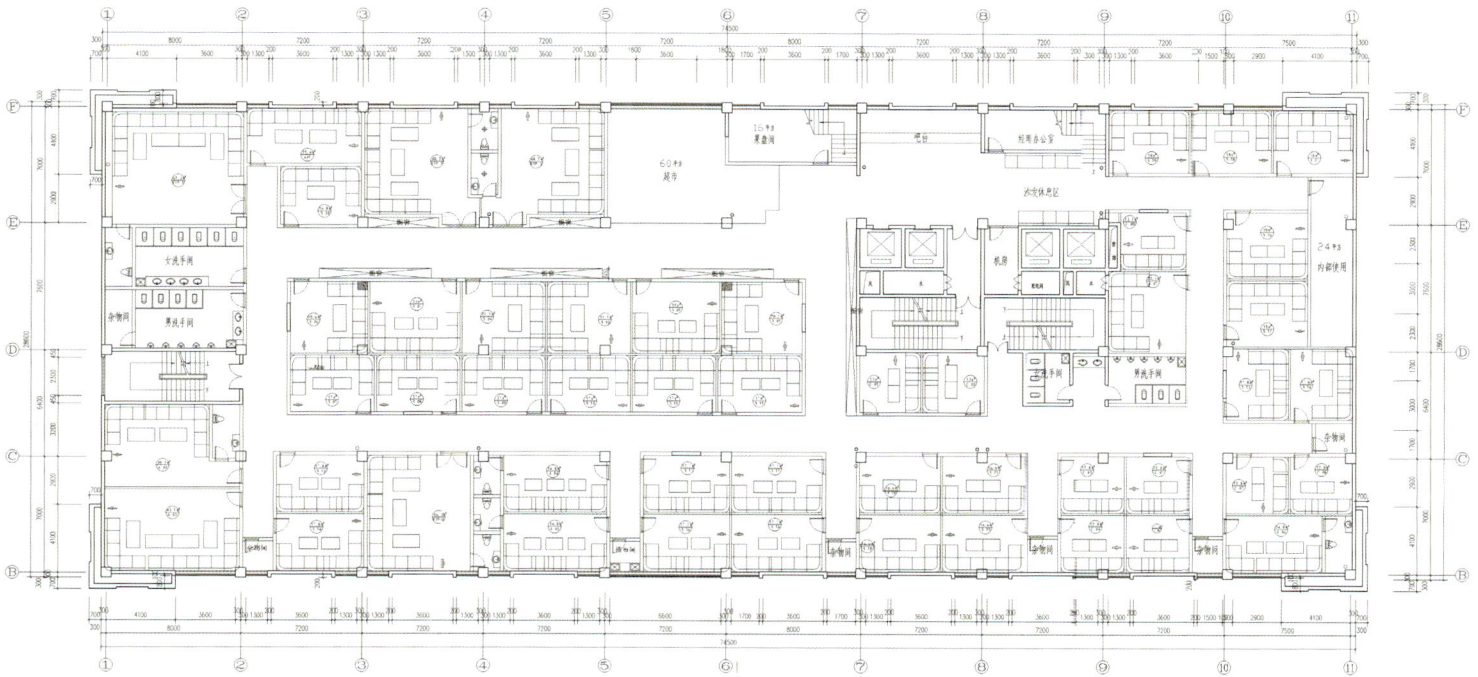

HAOMEN CLUB, GUANGZHOU

广州豪门俱乐部

—— 地点

广州

—— 面积

2400平方米

—— 设计师

孔志

—— 设计公司

广州构图室内装饰设计有限公司

—— 主要材料

银镜、黑金砂、邬钢、啡网石、布艺、马毛皮、

LED灯、镜钢、钛铜

A PRIVATE FEAST

Entertaining Club

DESIGN

私享盛宴：娱乐会所设计

The design of the project adopts the style of mix-and-match, which combines lots of design elements in a smooth way.

On the wall surfaces, silver mirrors and other reflective materials are massively used by the designer to extend the space. Classic scroll grass patterns are frequently seen on wall surfaces, cloth art and the patterns of the furniture, which not only shows the graceful taste but also enriches the shapes of the space.

Lotus patterns are frequently seen in the design of the lobby. The petal-shaped lighting effects in the center of the lobby and the double-petal lotus lamp on the ceiling fill the space with unseasonable warmth, plus the embellishment of the mirror steel, the blurring effects of LED lamps is displayed immediately and people get intoxicated in this blurring space of light and shadow, which is, so to speak, that it's not the wine that intoxicates but the drinker who gets himself drunk.

本案设计采用混搭风格，将诸多设计元素用圆滑的方式融合在一起。

墙面的处理上，设计师大量使用银镜等反光性材料来增加空间的延伸感，经典的卷草花纹也频繁出现在墙面、布艺及家具造型上，既显出优雅格调，又丰富了空间形态。

大厅的设计中频繁出现莲花造型，中央花瓣形的灯光处理及吊顶上的重瓣莲灯让空间充满着不合时宜的温馨气息，加上镜钢灯材料的晕染，LED的迷离效果顿时呈现，可谓是"酒不醉人人自醉"，让人沉醉在光影迷离的空间里。

三层平面布置图1:300

PRINCE KTV

王子KTV

— 地点
深圳
— 面积
4000平方米
— 设计师
吴羌
— 设计公司
深圳市吴羌室内装饰设计有限公司
— 主要材料
布衣、大理石、马赛克

In the hall, marble, glass, and the LED lights, which show up in a water-like posture, create a kind of vivid landscape, uplift the height of the space to its maximum, and bring people a feeling of disorder and confusion. People here freely get together and wait, whether to leave or enter, always in a pleasant mood.

The logo of "PRINCE" in English appears in the hall and extends to the design of the compartments. The lacquered glass and the mirror create lots of layered space. The red table explicitly displays the mark of "PRINCE", which enriches the spatial color and activates the spatial temperament. In the lighting treatment, a kind of relatively pure one is adopted, which avoids the colorful gorgeousness. It brings a feeling of cleanness and brightness and highlights the design theme of safety. health, pleasure and fashion.

大厅里，大理石与玻璃及LED灯以流水般的姿态营造出一道灵动的风景，最大限度地提升了空间层高，带给人错乱、迷离的感受。人们在这里自由聚集、等候，不管是离去还是进入，都会保持愉悦的心情。

大厅里出现的王子英文标识"PRINCE"延续到包厢设计中，烤漆玻璃与镜面材料打造出一个个富有层次感的空间。红色茶几上清晰地展示着"PRINCE"的标识，既丰富了空间色彩，又活跃了空间情绪。在灯光的处理中，选用比较纯粹的灯光，避免了五彩缤纷的绚烂，带给人干净、明快的感受，突出安全、健康、愉快、时尚的设计主旨。

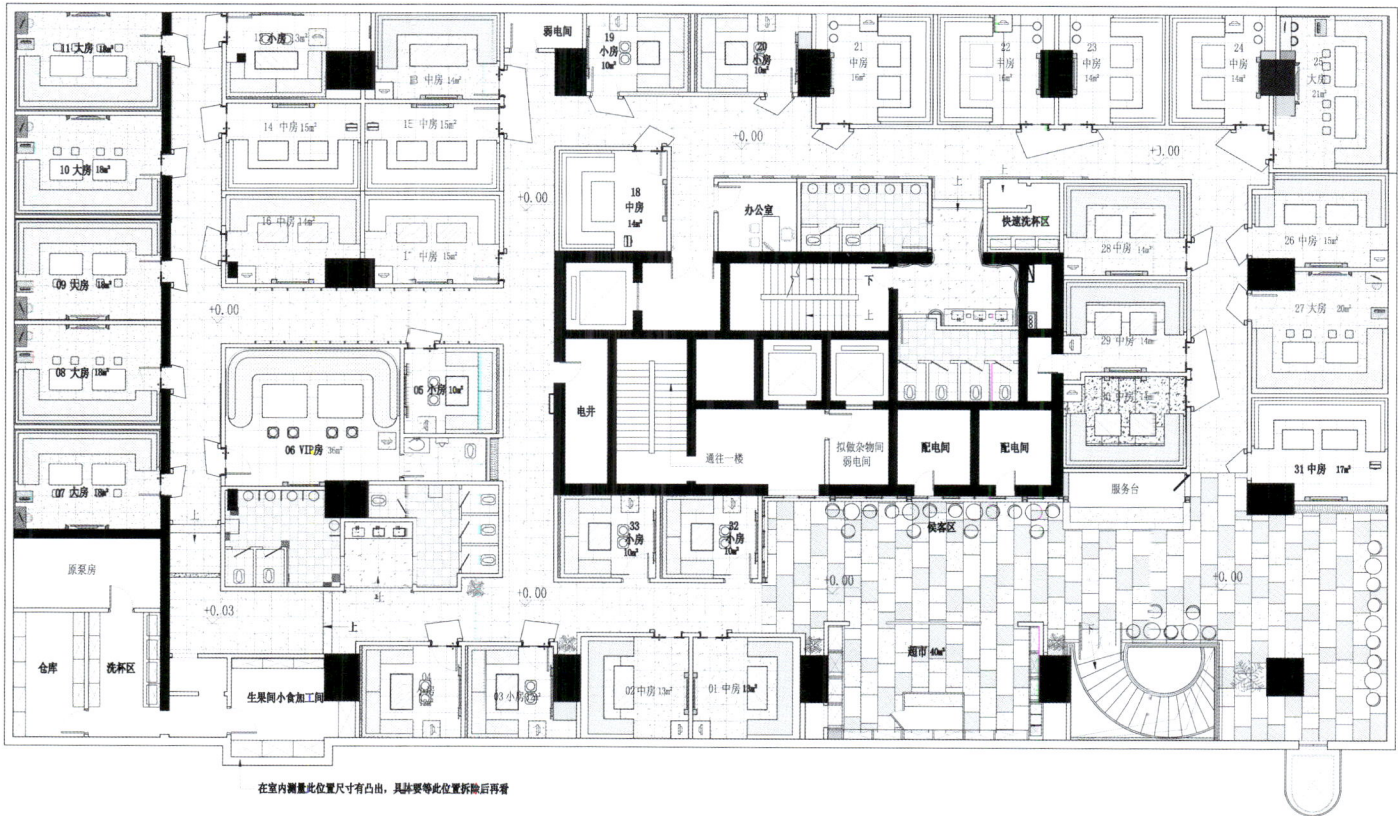

在室内测量此位置尺寸有凸出，具体要等此位置拆掉后再看

WUHAN CITY
GOLDEN AGE CLUB

武汉市金色年华俱乐部

— 地点
武汉
— 面积
约3000平方米
— 设计师
余良牧、关远传、胡娅
— 设计公司
广州市千浩室内装饰设计有限公司
— 主要材料
镜面不锈钢、玻璃镜面、有机玻璃造型、水晶挂
件、大理石、羊毛挂毯造型、光纤灯造型、LED
灯光造型

A PRIVATE FEAST
Entertaining Club
DESIGN

私享盛宴：娱乐会所设计

This project is located on the top floor of an old building. Rational plans and steel structure establishments make it a complete inside dining environment. In the banquet hall, symbols elaborately chosen by the designers are put on the wall decorations and the wavy wood partition modeling. A couple of birds bring the air of nature to the hall. A relatively brief design technique is employed in the interior design of the room and the contrast between materials and textures is highlighted to make the dining a double enjoyment of taste and vision. In addition, the lighting schemes combine the beautiful and elegant lamp ornaments with the LED lamps to provide a trace of splendor for the abundant modeling designs in the space, like coating the clay sculpture with a layer of gold, which immediately makes the entire space alive and vivid and the expressions in it also become varied and colorful. The intangible temperament not only provides people with the impulse to explore but also becomes a means to promote business.

本案位于一栋旧建筑的顶层，合理的规划和钢结构的搭建，使之成为一个完整的内部用餐环境。在宴会厅内，设计师精心挑选的符号被生动地运用到墙面装饰、波浪形的木隔断造型上，几只小鸟带给大厅不少的自然气息。房间内运用相对简约的设计手法，强调材质之间的对比，让进餐成为味觉和视觉的双重享受。此外，在灯光的处理上，用优雅美观的灯饰结合LED灯，为空间里丰富的造型添上一抹光彩，仿若给泥塑的雕像镀上一层金水，整个空间顿时鲜活生动起来，空间情绪也变得多姿多彩，捉摸不透的气质让人多了探奇的冲动，同样也可以作为促进生意的方式。

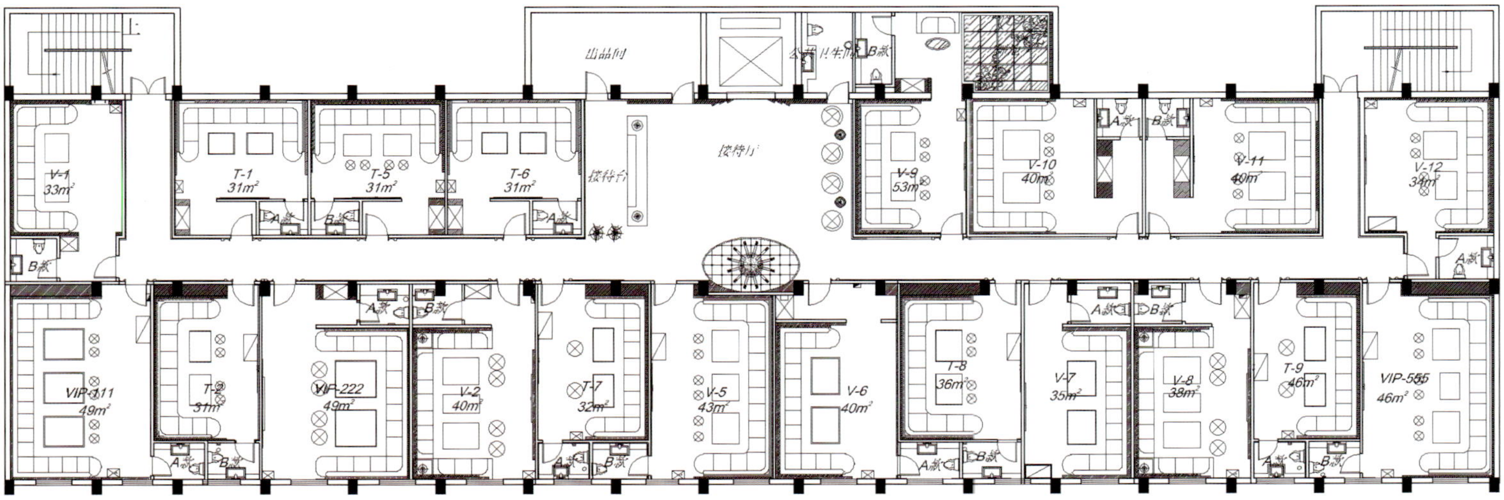
出品间
公共卫生间
接待厅
接待台
V-1
33m²
T-1
31m²
T-5
31m²
T-6
31m²
V-9
53m²
V-10
40m²
V-11
40m²
V-12
34m²
49m²
T-2
31m²
VIP-222
49m²
V-2
40m²
T-7
32m²
V-5
43m²
V-6
40m²
T-8
36m²
V-7
35m²
V-8
38m²
T-9
46m²
46m²

ARCADIA INTERNATIONAL CLUB, GUANGZHOU

广州世外桃源国际俱乐部

— 地点

广州

— 面积

3000平方米

— 设计师

郭为成、潘荔姗

— 设计公司

广州都市原点建筑装饰设计有限公司

A PRIVATE FEAST
Entertaining Club
DESIGN

私享盛宴，娱乐会所设计

People, who play different roles in the hectic life, urgently need a relaxing place to escape from their perplexing feelings and get intoxicated in the illusory and beautiful world, temporarily forgetting their worries. The designers complete this project with these ideas in mind aiming to build an arcadia where the memory of time and space fades just for getting intoxicated.

In here, luxury and dissipation cannot be found, but only the abundant, high-end spiritual life and material demands, which coexist together, can be truly satisfied in this elegant and gorgeous space. You can sit in a small hall, named Tailaisidian, in the club, open a bottle of Castel Merlot dry red wine or France royal dry red wine, listen to the ethereal music of shakuhachi, and watch the prosperous scene of the Guangzhou bustling street out of the window, and all these feel so close, and yet seem far away as well.

在忙碌的生活中扮演不同角色的人们，急需一个能够放松心情的地方，只为暂时逃开那些困扰自己的思绪，让自己沉醉在一个虚幻而美好的世界里，暂时忘掉忧愁。设计师怀着这样的想法来完成这个项目，意欲打造一个释放情绪的世外桃源，让人可以忘记时间和地点，只为沉醉。

在这里，看不到纸醉金迷的奢华，只有丰满奢侈的精神生活与物质需要，二者寄居共生，在这个优雅、华丽的空间里得到最真实的满足。你可以坐在世外桃源一个叫泰莱斯巅的小厅里，开一瓶卡斯特梅洛干红，或是法兰西贵族干红，听尺八如飘如缕的轻音，看窗外广州街道喧闹的繁华，而这一切，是这么近，又是那么远……

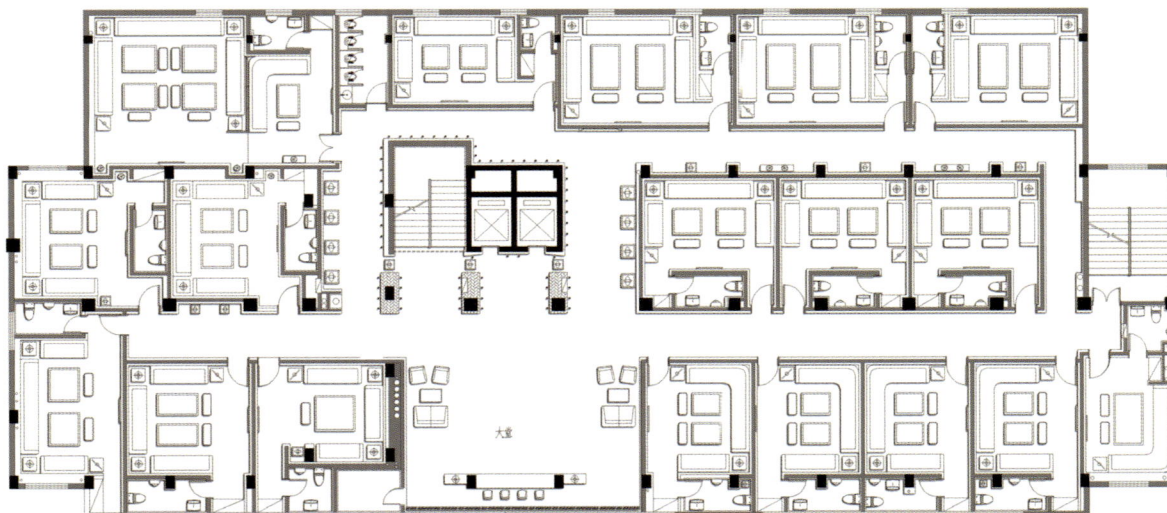

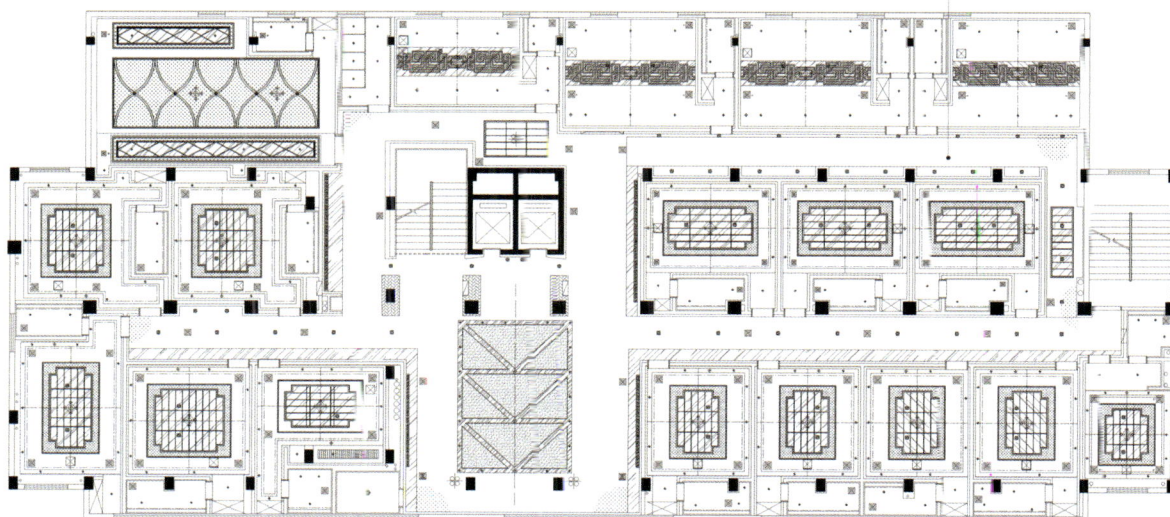

MERRIMENT RETAIL KTV

乐欢天量贩式KTV

— 面积
1340平方米
— 设计师
叶福宇
— 设计公司
深圳市菲尚装饰设计有限公司
— 主要材料
黑色抛光砖、灰镜

A PRIVATE FEAST
Entertaining Club
DESIGN

私享盛宴: 娱乐会所设计

This is a triangle-shaped space and the designer adopts an enclosed town-within-town layout. The medium-sized compartment rooms form the exterior walls of the space and the inner area has a straight-lined allocation, leaving the small triangles to be made as dining halls and bathrooms. Besides making the most of the space, materials and lighting are integrated to create a beautiful and bright space and a cordial and exciting atmosphere.

The designer chooses black polishing bricks, gray mirrors and the other dark color materials, combining with metal to build a space which is full of charms, leading the bewildered city people to experience a different kind of passion, relieving them of the tiredness overstocked in their hearts and providing a space to release their emotions. The lighting schemes integrate the form with the light. The beautifully designed crystal pendant lamp and the colorful LED lamps provide people with both visual and sensual satisfactions while brewing various space feelings, making it a charming place for people to linger on.

这是一个三角形空间，设计师采用"域中城"的围合形式来布局，中型包房构成空间外墙，中间区域采用直线型布局，剩下的小三角形也充分利用起来，设置饭堂和卫生间。在充分利用空间的同时，通过材料与灯光的结合，打造出一个绚丽闪亮的空间，营造热情喧闹的氛围。

设计师选用黑色抛光砖、灰镜等暗调材料，结合金属材质，营造了一个充满魅惑感的空间，带着在都市生活中迷惘的人群体验另一种热情，消除人们积压在心底的疲累，让人有一个释放情绪的空间。在灯光处理上，形式与光线相结合，拥有绚丽外观的水晶吊灯和缤纷的LED灯让人在视觉和感官上同时得到满足，同时酝酿出多种空间情绪，让空间充满魅惑色彩，让人流连忘返。

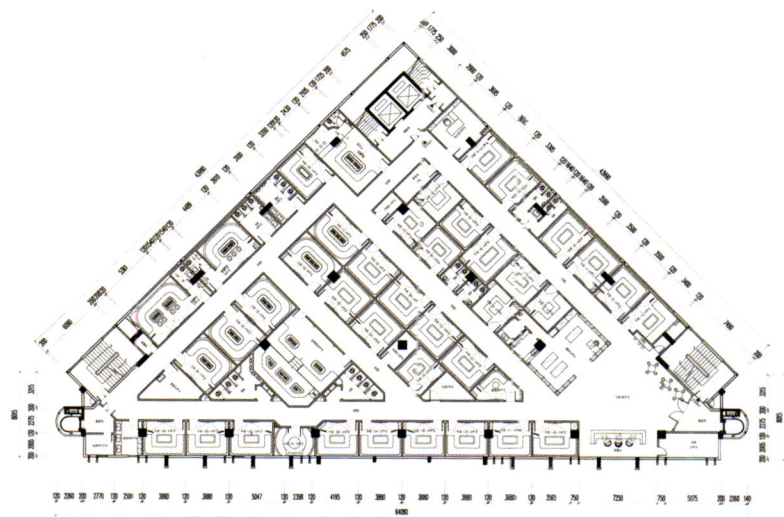